SPACE SCIENCE, EXPLORATION AND POLICIES

CIVIL EARTH OBSERVATIONS

NATIONAL STRATEGY AND PLAN

SPACE SCIENCE, EXPLORATION AND POLICIES

Additional books in this series can be found on Nova's website under the Series tab.

Additional e-books in this series can be found on Nova's website under the e-book tab.

SPACE SCIENCE, EXPLORATION AND POLICIES

CIVIL EARTH OBSERVATIONS

NATIONAL STRATEGY AND PLAN

BRENDON WALTERS
EDITOR

nova publishers
New York

For permission to use material from this book please contact us:
Telephone 631-231-7269; Fax 631-231-8175
Web Site: http://www.novapublishers.com

Additional color graphics may be available in the e-book version of this book.

Library of Congress Cataloging-in-Publication Data

ISBN: 978-1-63321-965-6

Published by Nova Science Publishers, Inc. † New York

CONTENTS

Preface		**vii**
Chapter 1	National Strategy for Civil Earth Observations *National Earth Observations Task Force*	**1**
Chapter 2	National Plan for Civil Earth Observations *United States Group on Earth Observations*	**67**
Index		**139**

PREFACE

Chapter 1 – Voluminous amounts of data and information come from a complex array of satellites, ocean buoys, stream gauges, human surveys, and other platforms for collecting what the scientific community calls Earth observations. These data are used every day to protect life and property and answer key questions about our planet.

This document establishes a framework for increasing the efficiency and effectiveness of the Nation's Earth-observation enterprise. Currently, 11 Federal departments and agencies engage in Earth observation activities, collecting volumes of important data about the Earth on an ongoing basis, using an array of sophisticated tools and systems. The Strategy outlines a process for evaluating and prioritizing Earth-observation investments according to their value to society in critical areas such as agriculture, global change, disasters, water resources, and weather.

Chapter 2 – The U.S. Government is the largest provider of environmental and Earth-system data in the world. These data are derived from observations of the Earth, which are used by Federal agencies and their partners to carry out their missions. These data form the foundation of services that protect human life, property, the economy, and national security, and they support research to foster scientific advances. The National Plan for Civil Earth Observations aims to maximize the value of observations collected by Federal agencies of the Earth's land surfaces, oceans, and atmosphere. The Plan is a blueprint for future Federal investments in and strategic partnerships to advance Earth observing systems. It lays out Federal priorities and supporting actions to manage Earth observation systems through routine assessments, improved data management, interagency planning, and international collaboration. These steps will ensure continued provision of, and enhanced access to, high-impact Earth observation data.

In: Civil Earth Observations
Editor: Brendon Walters
ISBN: 978-1-63321-965-6

Chapter 1

NATIONAL STRATEGY FOR CIVIL EARTH OBSERVATIONS*

National Earth Observations Task Force

EXECUTIVE SUMMARY

Each year, Federal agencies invest billions of dollars in civil Earth observations. Through these investments, the U.S. government ensures that the Nation's decision makers, businesses, first-responders, farmers, and a wide array of other stakeholders have the information they need about climate and weather, disaster events, land-use change, ecosystem health, natural resources, and many other characteristics of the planet. Taken together, Earth observations provide the indispensable foundation for meeting the Federal Government's long-term sustainability objectives and advancing U.S. social, environmental, and economic well-being.

As the Nation's Earth-observation capacity has grown, however, so has the complexity of this endeavor. Today, U.S. Earth-observation efforts are distributed among more than 100 programs under the purview of Federal agencies and non-Federal entities that both produce and use these data.[1] The National Strategy for Civil Earth Observations, developed in response to congressional direction,[2] is designed to maximize interagency coordination,

* This is an edited, reformatted and augmented version of a document issued by the National Science and Technology Council, Office of Science and Technology Policy, April 2013.

increase efficiency and efficacy of future Earth-observation efforts, and promote environmental and economic sustainability.

The National Strategy includes two main elements:

- A policy framework and method for Federal Government assessment of Earth observations (data and products derived from Earth-observing systems) that defines societal benefit areas, evaluates information products produced for those areas, identifies critical data streams that support those products, and prioritizes the observing systems on which those data streams depend, taking account of both present capabilities and anticipated needs and technologies.
- Data-management guidelines that advance data-management frameworks and improve information-delivery systems for Earth-observation data.

A third element—a National Plan for Civil Earth Observations, based on the results of the assessment and that takes fiscal and programmatic constraints into account—is to be published as a supplement to the President's budget for fiscal year (FY) 2014.

This National Strategy pertains to observations and data for civilian-use purposes. It strives toward openness and integration of data from Federal and non-Federal sources, including unclassified products derived from classified sources, while explicitly recognizing that certain sources of data may require greater oversight, protection, and potential restriction due to national-security, privacy, and confidentiality considerations.

The National Strategy provides a method of assessing Earth-observing systems by defined societal benefit areas—a key prerequisite to crafting a National Plan for Civil Earth Observations. This organization around benefit areas is designed to consider the relative impact of observations from multiple sources, as well as the need for measurement continuity for societal benefits that depend on sustained, long-term, and accurate measurements.

While Earth observations are typically produced for a specific purpose, they are often found to be useful for additional purposes not foreseen during the development of the observation system. Earth-observation data should be managed and preserved such that both anticipated and unanticipated users can find, evaluate, understand, and use the data in new ways. This National Strategy, therefore, sets out a comprehensive data-management framework as an integral element of current and future Earth-observation planning, to ensure effective and continuing access to Earth observations by the full range of potential users and beneficiaries.

The National Plan for Civil Earth Observations, once completed for FY 2014, is intended to be updated every 3 years to inform Federal investments in civil Earth-observation systems. This approach of routine assessment, improved data management, and coordinated planning is designed to enable stable, continuous, and coordinated Earth-observation capabilities for the benefit of society.

1. OVERVIEW AND POLICY FRAMEWORK

1.1. Introduction

Federal investments in civil Earth observations provide critical societal benefits by enabling scientists, decision-makers, and citizens to advance the drivers of U.S. economic, social, and environmental well-being. Such investments allow the deployment of renewable energy technologies and other sustainable practices such as integrated water-resource management, coastal and marine planning, climate-adaptation strategies, and sustainable land use. Recent Earth-system science research has also revealed important new knowledge that results directly from closing gaps in critical time-series observations and improving observations of complex Earth-system processes, including human activity.

Without routine and comprehensive Earth observations, the Nation risks not having adequate informa-tion to respond to crises and to evaluate potentially harmful long-term trends. For example:

- During natural and human-induced disasters, both existing observational data and real-time monitoring are needed to develop effective response measures that mitigate loss of life and property.
- A lack of long-term and continuous observational data undermines the credibility of environ-mental data records.
- Sudden loss of, or unexpected data gaps in, routine observations can dramatically affect fore-casting and modeling accuracy. This can contribute to poor decision-making with substantial economic and social costs. Certain data gaps are irreplaceable, inhibiting the ability to link past and future measurements.

As such, Earth observations provide the basic information necessary for meeting the Federal Government's long-term sustainability objectives, which include: (1) encouraging the use of sustain-ability science as a mechanism to integrate human and natural sciences, ecosystem science, ecosystem recovery and restoration work, land-use and watershed considerations, and related areas for environ-mental and natural resource R&D, policy, and management; and (2) supporting science and technology for sustainability to promote sustainable products, processes, and systems.

The myriad Earth observations made today are distributed among more than 100 programs under the purview of Federal agencies and non-Federal entities that both produce and use Earth-observation data.[3] The magnitude and complexity of Federal-agency investments in civil Earth-observation activities require a regular, coordinated effort to review, assess, and optimize those investments. Achieving a greater degree of interagency coordination will improve the efficiency and efficacy of future Earth-observation efforts and promote environmental and economic sustainability.

1.2. Definition of Earth Observations

For the purposes of this document, the term "Earth observations" refers to data and products derived from Earth-observing systems and surveys. "Observing systems" refers to one or more sensing elements that directly or indirectly collect observations of the Earth, measure environmental parameters, or survey biological or other Earth resources (land surface, biosphere, solid Earth, atmosphere, and oceans).[4] Sensing elements may be deployed as individual sensors or in constellations or networks, and may include instrumentation or human elements. Observing-system platforms may be mobile or fixed and are space-based, airborne, terrestrial, freshwater, or marine-based. Observing systems increasingly consist of integrated platforms that support remotely sensed, in-situ, and human observations.

1.3. Guidelines for the National Plan for Civil Earth Observations

In response to Congressional direction, the National Strategy for Civil Earth Observations establishes guidelines, processes, and criteria for the development of a National Plan for Civil Earth Observations to guide Federal investments in civil Earth-observation systems. The purpose of the plan will be to enable stable, continuous, and coordinated global Earth observation capabilities for the benefit of society. The National Plan will be informed by an assessment process for determining priorities, which will use an approach that explicitly links goals and strategies to resource availability and optimized allocations.[5]

The National Plan, targeted for initial publication for FY 2014, will be updated every 3 years to ensure it reflects revised observational requirements, new and improved technologies, domestic and international developments, and revised budget outlooks. A 10-year horizon will be considered for the first National Plan and in each update to incorporate the evolution of Earth-observing needs, assess near- and long-term information requirements, and identify methods to better integrate the U.S. portfolio of systems, networks, and platforms necessary to meet those requirements. This includes the data-management and information-delivery mechanisms integral to all Earth-observation systems. In addition, the National Strategy envisions close collaboration with the modeling community and the private sector to ensure that (1) observations are integrated with Earth-system models and that the observation needs of the modeling community are addressed, and (2) data-management practices are established in coordination with existing community (non-Federal) data standards, where applicable.

The National Plan will be developed in accord with the following principles:

- **Information for Societal Benefit**: Identify information requirements and measurements needed to maximize societal benefits, in the areas described in Section 1.4 of this National Strategy. Specific attention will be given to observation of complex Earth-system processes (including coupled human and natural systems); observational requirements for Earth-system models; and acquisition and timely delivery of requisite and relevant information to scientists, service providers, on-scene emergency coordinators, public and private decision-makers, and the public. See Appendix B for elaboration of the questions that inform societal benefit.
- **Integrated Portfolio Management**: Manage the Federal Government's Earth-observation investments as a prioritized portfolio that integrates within and across the areas described in Section 1.4; integrate information requirements and measurements; coordinate federally operated systems, networks, platforms, and programs;

identify reliance on non-federally operated systems; optimize investments among different platforms; and optimize and balance research-data collection with operational monitoring.

- **Federal Collaboration**: Promote and facilitate effective collaboration among diverse Federal Earth-observation activities, and clarify observing and associated life-cycle data-management responsibilities, including metrics for responsiveness and timeliness.
- **Interoperable Systems and Access**: Advance the integration and harmonization of data at the sensor, platform, data-processing, and application levels to ensure that data are comparable independent of sensor, platform, and organization; promote a service-oriented, standards-based technical architecture to provide interfaces to Earth-observing systems and their associated data-management systems, ensuring that users can both find and visualize the data; establish a user-driven policy and architecture for access to and archiving of federally funded observations (or other data by agreement) to ensure that they are widely available and accessible in a timely, efficient, and user-friendly manner.
- **Domestic and International Partnerships**: Identify critical dependencies and successful partnerships with non-Federal observing networks and achieve a greater degree of interagency coordination to improve the efficiency, efficacy, and costs of leveraging the observation net-works of non-Federal entities.[6] See Appendix C for details.
- **U.S. Technical and Management Leadership**: Advance U.S. technical leadership in measurement and monitoring technologies by investing in a range of research and development that supports Earth observations, including the development of technologies, sensors, applications, and alternative approaches to recognize, prioritize, and appropriately respond to emerging information needs.
- **National-Security Assets**: Increase the use of national-security assets for civil purposes, where appropriate and in coordination with the Civil Applications Committee. Take the availability of these data into account when planning new missions, while maintaining protections for national security, personal privacy, and civil liberties.[7]

1.4. Observations and Information to Benefit Society

This Strategy establishes a process to evaluate Earth-observing systems based on the information products and data streams they support in defined societal benefit areas (SBAs). This approach was adopted by the National Science and Technology Council Committee on Environment, Natural Resources, and Sustainability in February of 2012, is consistent with the Federal Government's sustainability objectives, and aligns with international agreements and prior interagency work in this arena.[8] The first National Plan for Civil Earth Observations will be organized around 12 SBAs (listed alphabetically) and the reference measurements that underpin them:

- **Agriculture and Forestry**: Supporting sustainable agriculture and forestry.
- **Biodiversity**: Understanding and conserving biodiversity.
- **Climate**: Understanding, assessing, predicting, mitigating, and adapting to climate variability and related global change.
- **Disasters**: Reducing loss of life, property, and ecosystem damage from natural and human-induced disasters.
- **Ecosystems (Terrestrial and Freshwater)**: Improving the management and protection of terrestrial and freshwater ecosystems.
- **Energy and Mineral Resources**: Improving the identification and management of energy and mineral resources.
- **Human Health**: Understanding environmental factors affecting human health and well-being.
- **Ocean and Coastal Resources and Ecosystems**: Understanding and protecting ocean, coastal, and Great Lakes populations and resources, including fisheries, aquaculture, and marine ecosystems.
- **Space Weather**: Understanding, assessing, predicting, and mitigating the effects of space weather on technological systems, including satellites, power grids, communications, and navigation.
- **Transportation**: Improving the safety and efficiency of all modes of transportation, including air, highway, railway, and marine.
- **Water Resources**: Improving water resource management through better understanding and monitoring of the water cycle.
- **Weather**: Improving weather information, forecasting, and warning.

- **Reference Measurements**: Improving reference measurements—the underpinnings of all the SBAs—such as geodesy, bathymetry, topography, geolocation, and the fundamental measurement systems and standards supporting them.

These SBAs are interconnected and include scientific research, economic activities, and environmental and social domains. Multiple SBAs share observational requirements, and societal benefits accrue from Earth observations that inform scientific research, policy, and decision-making. Earth observations also support critical government functions, such as the continuity of national government and the protection of life and property. The SBAs are described in greater detail in Appendix B.

2. Assessment Method and Plan Development

2.1. Introduction

This chapter describes the assessment method for the development of a National Plan for Civil Earth Observations and subsequent triennial updates. The overall process is designed to be accomplished in two phases:

- **Phase 1: Portfolio Assessment**. The foundational element for the National Plan is an internal assessment of existing and planned observing systems in providing environmental observations to benefit society across the SBAs. The assessment evaluates the current portfolio and makes recommendations, for internal Federal use, concerning continuity, fulfillment, and advancement of required measurements over a 10-year planning period,[9] taking account of existing capabilities and anticipated needs and technologies.
- **Phase 2: National Plan for Civil Earth Observations**. Using the portfolio assessment, the Executive Office of the President (EOP), in close coordination with agency leadership, will develop a cross-agency National Plan consistent with the President's budget for specific Earth observation systems, platforms, and networks. This plan will be published as a supplement to the President's budget.

2.2. Phase 1: Portfolio Assessment

The portfolio assessment process is designed to review the impact of observations from multiple sources. The process examines measurement continuity for societal benefits that depend on sustained, long-term, and accurate measurements.

The first portfolio assessment was carried out in 2012 by an interagency Assessment Working Group appointed by the NEO Task Force, supported by expert teams for each societal benefit area to aid the EOP in establishing Earth observation priorities for the Nation. In evaluating existing systems and making recommendations for the future, the assessment process considered, among other criteria, the following characteristics of observing systems:

- Impact on a single SBA or multiple SBAs.
- Ability to complement other observational systems, including international, commercial, civil, and non-civil systems.
- Degree of risk, including readiness (technical, resources, people).
- Mitigation of risk and strategic redundancy (backup of other critical systems).

The Assessment Working Group began by refining the assessment method and prioritization criteria presented in this National Strategy. The Working Group drew from experts who are familiar with the use of remote sensing, in-situ, and other Earth observation techniques. The group also drew on current and potential users who focus on the development of decision-essential science and tools using the data.

2.2.1. SBA Assessment Teams

To ensure that the right expertise was available and engaged throughout the process, technical experts were identified to lead and support assessment teams for each SBA. The Assessment Working Group worked with subcommittees of the National Science and Technology Council's (NSTC) Committee on Environmental, Natural Resources, and Sustainability (CENRS) to identify lead and contributing experts for each team. The Assessment Working Group provided centralized guidance and oversight to ensure that each of the SBA assessments were completed to the same level of detail and that each team applies the assessment criteria uniformly across the SBAs.

2.2.2. SBA Assessment Process

The SBA assessment teams used the process outlined below to complete an assessment and evaluate the impact of observing on each of the SBAs.

1. ***Identification and Definition of SBA Subareas***. SBAs transcend individual agency needs and mandates, and describe broad societal, scientific, and economic goals. Because SBAs are broad, the first step was to identify major subareas within each SBA. This facilitated broad coverage as well as provided a framework for identifying the needed expertise.
2. ***Identification of Key Objectives.*** SBA teams identified key objectives for each subarea. Key objectives are those activities, products, services, or desired end states which rely in whole or in part on Earth-observation data to provide societal benefit. The objectives clearly identified what is being achieved using Earth observations.
3. ***Identification of the Current Observing System Portfolio.*** SBA teams identified data sources (observing systems providing direct observations) and tools (models, products, etc.) that are required to achieve the key objectives. Where products or models were identified, an additional step was necessary to determine the observing systems providing data to the model or used to produce the product. This value chain analysis was repeated until all observing system inputs supporting the objective are identified (see Figure 1). The final result is the list of observing systems currently contributing to achieving benefits in the 12 SBAs plus the reference measurements that underpin each.
4. ***Assessment of Current Observing System Portfolio.*** The Assessment Working Group will provided standardized guidance to the SBA teams, defining the method for determining the relative impact and performance of the observing system portfolio identified in Step 3. The method also included an evaluation of the current performance level for each objective, sub-area and SBA, which formed the baseline for evaluating and prioritizing gaps in step 5. The result of this step was an impact-ranking of the current portfolio of observing systems.
5. ***Development of Recommendations for the National Plan for Civil Earth Observations.*** Starting with the impact ranking, the SBA teams developed narrative recommendations, described in terms of maintaining continuity, mitigating risks and gaps, and identifying opportunities to capitalize on new or improved technologies that add

capability, replace existing capabilities with more efficient cost-effective methods, plan for long-term data-set development, and leverage partnerships.

2.2.3. Integrated Assessment across SBAs

Once the individual SBA assessments were completed, the final activity in Phase 1 was for the Assessment Working Group to identify priorities for observational technologies and capabilities according to their contribution to measurement needs over the long-term planning horizon. The results of this assessment were provided as internal, deliberative input to the EOP for the formulation of the National Plan for Civil Earth Observations.

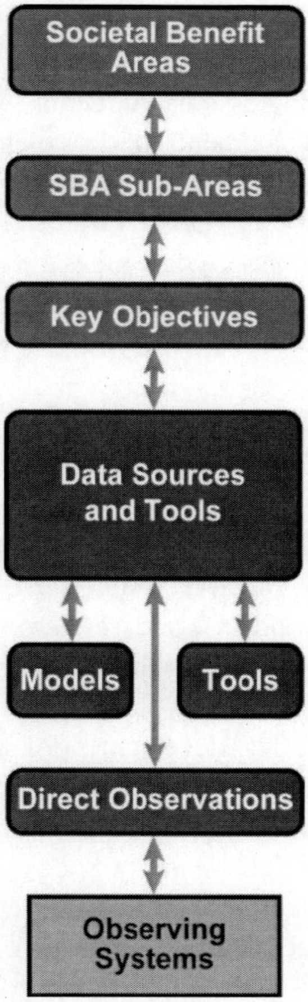

Figure 1. Value chain analysis.

2.3. Phase 2: National Plan for Civil Earth Observations

The second phase is the development of the National Plan for Civil Earth Observations, which will be informed by the internal assessment as well as other external sources, such as the many reports of the National Academies relating to Earth observations. The National Plan will be published as a supplement to the President's budget. Building on the guidelines outlined in Section 1.3, the plan will take into account the following major considerations:

- **Coordinating Earth Observations from Space:** Given the large Federal investment in Earth observation satellite development and operation, strengthened planning about the require-ments for space-based observation is necessary, as is investigating alternative technologies and approaches to meeting these requirements. In accordance with congressional direction, specific attention will be given to the development of a plan for sustained climate measurements from space.
- **Sustaining Long-Term Observations:** The National Plan will consider issues and options concerning the sustainability of civil Earth observations from demonstration and research programs, on all platforms, to long-term data-acquisition programs. (See Chapter 3 for more detail).
- **Delivering Integrated Information:** The Nation's investment in data-management and information-delivery systems is an integral part of the National Earth observation infrastructure. This National Strategy respects the "presumption of openness" called for in the 2009 Open Government Directive. Full, open, and timely sharing and exchange of data, products, and services through machine-readable formats and Application Programming Interfaces (APIs) will enable users to effectively and rapidly transform data into useful information products for society. The development of widely accepted standards, principles, and guidelines for data management, dissemination, and quality will encourage and facilitate the integration of diverse U.S. Earth observations for the benefit of society, and it will contribute to existing and coordination efforts. This National Strategy adheres to the principle that access to data-managed or paid for using Federal funds should be open to the public as soon as possible after collection, in a nondiscriminatory manner, and generally free of charge. Agencies

may set user charges in a manner consistent with Office of Management and Budget (OMB) Circular A-130.[10]

- **Affordability**: Cost considerations, including development, deployment, operations, and maintenance costs year and integrated costs of missions and programs over multiple years.
- **Identifying Common Needs and Sharing Resources:** Data needs across SBAs will be compared to identify common requirements and opportunities to share observing and data-management systems resources.
- **Interdependency of Federal and Non-Federal Observing Systems:** Observational networks funded in part by the Federal Government may also contain components owned by non-Federal entities. In some cases, these systems may be more cost effective than wholly Federal observational networks, but they can also require complex funding and management arrangements, which involve risk because quality control and continuation of these observations is not under the direct control of the Federal Government. Governance of these systems must be accomplished through coordinated policies for inventory control, data ownership and sharing, and public-private partnerships. Similarly, observational networks that are supported by more than one agency (e.g., ground-based measurements needed to support space-based measurements) pose special challenges to continuity and will need to be coordinated.
- **Engaging the Private Sector and the Public:** Mechanisms for engaging industry, non-governmental organizations, and other stakeholders in Earth observation will be developed or enhanced to take advantage of the competencies and contributions of stakeholders and the potential benefits of increased U.S. competitiveness in the private sector.
- **Improving User Engagement:** Mechanisms for engaging user communities will be developed to cultivate wider application and use of Earth observations.
- **Leveraging National Security Assets:** Existing mechanisms for increasing the use of national security assets for civil purposes will be strengthened, while maintaining protections for national security, personal privacy, and civil liberties.

Building on the assessment of Earth observation systems and needs developed in Phase 1, the Executive Office of the President (EOP), in close coordination with agency leadership, will develop the National Plan. Consistent with congressional direction, the National Plan will address agency roles and responsibilities in Earth observations and will take account of known budgetary and programmatic constraints.

The EOP will complete a first draft of the National Plan for Civil Earth Observations. The final draft will be reviewed by CENRS and published by the EOP as a supplement to the President's FY 2014 Budget. As required by Congress, the plan will be updated every 3 years, including a process for independent external advisory input.

3. Data Management and Delivery

3.1. Introduction

This chapter presents a framework for the delivery, quality, and preservation of Earth-observation data, including data-management principles, guidance, approaches, architectural considerations, standardization, and evaluation. This framework incorporates concepts and recommendations from recent studies by the National Research Council, the Interagency Working Group on Digital Data (IWGDD), the United States Group on Earth Observations (USGEO), the United States Chief Information Officer, the strategic plans of the participating agencies, and Open Government initiatives such as Data.gov.[11]

While Earth observations are typically produced for a specific purpose, they are often found to be useful for additional purposes not foreseen during the development of the observation system. It is important that Earth-observation data be managed and preserved such that both anticipated and unanticipated users can find, evaluate, understand, and utilize the data in new ways. The wide range of scientific and observation efforts of Federal agencies, and the resulting magnitude of data collections and diversity of data types, requires an approach that is broadly applicable yet able to be tailored to particular needs. Specifically, the National Plan will take into account the many observation and data-management initiatives currently under development (see textbox on the following page).

This document sets out a comprehensive data-management framework of transparent, evolvable, extensible policies and associated management and organizational structures to provide effective and continuing access to Earth observations. This framework sets expectations and requirements for Federal agencies involved in the collection, processing, stewardship, and dissemination of Earth-observation data. The data policies of entities outside the Federal Government are beyond the scope of this document. The framework described here, however, applies to the use of all data by Federal agencies, regardless of source. The goal is to (1) maximize the likelihood that Earth observations are available and disseminated in a timely and usable manner, (2) facilitate the transformation of observations into useful information through the use of open, machine readable-formats and APIs, and (3) encourage the development and use of uniform tools and practices across Federal agencies for the handling of Earth-observation data to increase interoperability. The framework should guide and inform the development and implementation of agency- and program-specific data-management plans.

Toward those ends, Federal agencies should collect, maintain, disseminate, and preserve Earth observations so that the resulting data and products can be readily shared with other Federal agencies and non-Federal users, promoting data integration between sources. Agencies should ensure that Earth-observation data and products are included on agency record schedules that have been approved by the appropriate authorities.[12] These activities will adhere to appropriate standards and be conducted in accordance with existing regulations.

DATA AND INFORMATION MANAGEMENT INITIATIVES

A comprehensive data-management framework, as outlined in this strategy, will be necessary to support the following initiatives, among others:

GCIS: The Global Change Information System (GCIS) has the immediate priority of meeting the needs of the National Climate Assessment (NCA). The goal of the GCIS will be to integrate multiple agency components into an interagency system that allows easy access to usable climate services.

Ocean.data.gov: Created by the National Ocean Council to support Coastal and Marine Spatial Planning.

U.S. IOOS DMAC: The Data Management and Communications (DMAC) subsystem of the IOOS is overseen by the Subcommittee on Ocean Science and Technology's (SOST) Integrated Ocean Observations Committee.

EcoINFORMA: A recommendation by the President's Council of Advisors on Science and Technology (PCAST) was for an Ecoinformatics-based Open Resources and Machine Accessibility initiative to better enable use of bioinformatics data.

Biodiversity Monitoring: PCAST also recommended a data gaps and priorities assessment of Federal and regional ecological monitoring systems, accompanied by integrating existing monitoring networks.

Data.gov and OpenEI: Energy resource and consumption datasets, tools, and applications in open government platforms, such as Data.gov/Energy and OpenEI.org, already aid in identifying climate change mitigation opportunities.

WaterML: Water Markup Language (ML) is an informatics initiative of the CENRS Subcommittee on Water Availability and Quality that provides a systematic way to access water information from point observation sites.

LCA database: The Life Cycle Assessment (LCA) database is the United States Department of Agriculture's new initiative to incorporate environmental and crop data into a life cycle assessment of manufactured products and other goods and services.

DOI Initiatives: Information compiled and collected through the Landscape Conservation Cooperatives, and Climate Science Centers established under the DOI's climate change science initiative, will be part of a national and, ultimately, international network.

OFCM: The Office of the Federal Coordinator for Meteorology (OFCM) Committee for Integrated Observing System coordinates weather and climate observing networks to create a *network of networks*.

All of these information initiatives are interlinked and comport with Data.gov activities under the Open Government Initiative, and the Digital Government Strategy, including the merger of the Federal Geographic Data Committee's Geospatial OneStop into Data.gov.

Federal agencies should continue efforts to encourage data openness, data sharing, and increased data access in the international system, in particular by advancing the data sharing principles of the Global Earth Observation System of Systems (GEOSS).[13]

The following definitions apply to the key concepts in this chapter:

- **Data:** For the purposes of this discussion, we use the terms "data" and "Earth observations" interchangeably to mean geo-referenced digital information about Earth, including the observations, metadata,[14] imagery, derived products, data-processing algorithms (including computer source code and its documentation), and forecasts and analyses produced by computer models. Non-digital data, published papers, preserved geological or biological samples, or other media that have not been digitized are not included in this definition and are outside the scope of this chapter.
- **Data-Management Framework:** The policies, requirements, activities, and technical considerations relevant to data produced or handled by an agency or program.
- **Data Life Cycle:** All the activities that affect data before and during its lifetime, including planning for and producing, managing, disseminating, using, and disposing of the data.[15]
- **Data-Management Life-Cycle Process:** An approach to managing the flow of an information system's data throughout the life cycle of the data.
- **Data Durability:** Durable data are those that are managed to ensure lasting value and usability for both their original purpose and new applications, thus providing maximum utility for scientific understanding, decision-making processes, and public access today and in the future.
- **Data Delivery**: The delivery or hosting of Earth observations so that the widest possible number of users may access the data, agnostic of platform. The development of specific information service delivery mechanisms, such as portals, for specific communities is beyond the scope of this document.

The concepts in this National Strategy pertain to data for civilian-use purposes; they reflect the goal of greater openness and integration of data from Federal and non-Federal sources, including unclassified derived products from classified sources. This Strategy explicitly recognizes that certain sources of data may require greater oversight, protection, and potential restriction due to national security, privacy, and confidentiality considerations.

3.2. Data-Management Framework

The data-management framework is founded on the core concept of the data life cycle, defined below, and several crosscutting themes, including principles, guidance, architecture, standards, and evaluation, which are also discussed below. It is important to note that the crosscutting themes can be applied to the management of all data sets, and there are also separate lifecycles for each data set.

3.2.1. Data Life Cycle

The core of the data-management framework is the *data life cycle*, which can be divided into three activities as shown in Figure 2:

- Planning and production, which includes all activities up to, and including, the moment that observations are collected.
- Data management, which includes all activities related to collecting, storing, verifying, documenting, advertising, distributing, disposal, and preserving the data.
- Usage, which includes all activities performed by the consumer of the data.

These data life cycle activities are discussed in greater detail in Appendix D.

A life-cycle data-management process promotes durability of data; ensures that users are able to easily discover and access data, understand their context, and use them both for their intended purpose and in novel ways to aid research and decision-making; and ensures the durability of any derived data by linking their context to that of the original data upon which they are based.

Data life cycle process steps are interdependent. For example, inadequate documentation at an early stage can prevent later use; generating products from original data may yield new derived data that must also be collected and managed; and user feedback regarding data may change or augment documentation relating to the data. Because data may go through multiple cycles of use and reuse by different entities for different purposes, effective management of each step and coordination across steps in the life cycle are required to ensure that data are preserved and can be accessed and used efficiently and for new applications.

In the proposed process, data and associated metadata are acquired and incorporated into Federal data archives, where the data are evaluated and validated, supplemental data may be added, and data and data sets are maintained through a stewardship process that may involve both Federal experts and users. These data may further be integrated into new or enhanced data sets or products that reside in the repository or archive for use. Data and products, as well as metadata and standards, should be continually reviewed and improved throughout the data-management life cycle.

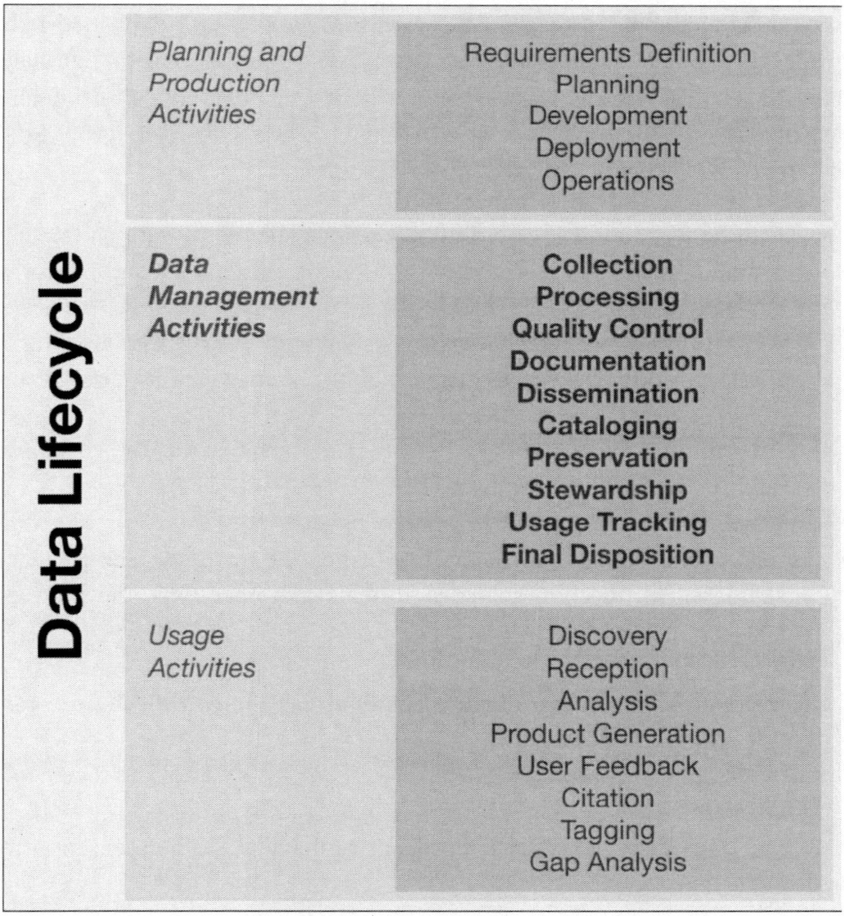

Figure 2. Activities in the data life cycle. Data-management activities, emphasized in this chapter, are shown in bold.

The resulting system must provide consistent, efficient, open, and nondiscriminatory access across multiple Federal data archives and repositories. These guidelines must also take into account access to both new and legacy systems, preserving compatibility when possible to allow users access to, and discovery of, archived Earth observations.

3.2.2. Principles

The data-management framework provides a set of basic principles under which relevant Federal agencies will develop their own data-management plans and requirements. The principles are summarized and explained here, and relevant guidance that supports these principles is provided in later sections of the chapter. These principles are elaborated in Appendix D and are in accordance with United States Global Change Research Program (USGCRP) data principles (see Annex B).

- **Full and open access:** Earth observations should be fully and openly available to all users promptly, in a nondiscriminatory and platform-agnostic manner, and generally free of charge wherever possible. Agencies may set user charges consistent with OMB Circular A-130.
- **Preservation:** Earth observations should be managed as an asset and preserved for future use.
- **Information data quality:** Earth observations should be of known quality and fully documented.
- **Ease of use:** Earth observations should be easily discoverable, searchable, and accessible online using interoperable services and standardized, machine-readable formats that encourage the broadest possible use.

These principles apply to all Federal Earth observations, with case-by-case exceptions for particular data sets.

3.2.3. Guidance

All civilian Federal agencies that produce or manage Earth observations or related data should:

- Establish a publicly available open data policy and a supporting data-management approach that addresses all aspects of the data life cycle, including open publishing of data, and metadata and quality-assurance management.

- Solicit and acknowledge stakeholder requirements and feedback with an understanding of the data use in advancing environmental, economic, and societal sustainability.
- Make data and services available in ways that support a web-based, service-oriented architecture (SOA) approach as described in Exchanging Data for Societal Benefit: An IEOS Web Services Architecture (USGEO 2008).[1620]
- Provide nondiscriminatory data access and dissemination to interested users.
- Determine what data require long-term preservation and how preservation will be accomplished in conjunction with National Archives and Records Administration. Whenever possible, coordinate observation and data-management systems with appropriate archives early in the life cycle.
- Support data durability by providing assured delivery to an archive, secure storage, secondary offsite storage, and ongoing data stewardship to ensure continued accessibility and readability.
- Implement interactive discovery and access mechanisms for data and web services compatible with those of other agencies, including Data.gov and GEOSS, and other non-Federal users.
- Enable integration and interoperability with broadly used data-analysis software and decision-support tools. Demonstrate these approaches for derived products and delivery.
- Acknowledge that data-management and information delivery services are fundamental, integral aspects of any Earth observation endeavor, and provide sufficient resources in agency funding submissions for data-management and information delivery.

The data-management framework does not impose a one-size-fits-all plan. Given the wide variety of agency missions and data types, it is not reasonable to expect that a single plan can apply to multiple agencies or even throughout a single agency. Agencies have the freedom to create flexible approaches or plans that conform to specific operational needs of each program or data set, account for Earth observations through their life cycles, and allow for special cases. Agencies should develop capacity to assist in developing life-cycle data-management plans in accordance with the principles laid out in this framework.

In addition to agency activities, robust, focused, and sustained interagency coordination is needed to foster compatibility of data and systems from different agencies; to reduce costs by sharing software, standards, and best practices; to pool resources in acquisition of data-management technology or enhancements; and to avoid duplicative or conflicting methods and acquisitions.

3.2.4. Architecture

Earth-observation data exist in and are used across a number of heterogeneous platforms managed by multiple independent operators. These diverse systems cannot be subsumed into any single master system, but can be harmonized and connected through the use of interoperability standards and open publishing of data. Federal agencies should pursue an information system-of-systems approach to leverage and harmonize multiple legacy, modern, and future component systems that have evolved separately and are managed independently. An Earth observation information system of systems is a collection of agency- or project-specific information systems that are loosely coupled to deliver information and services to users while enabling each agency or program to remain the steward of its own information. This approach is enabled by the Internet and is the practical basis of GEOSS, referenced in Chapter 1.

A fundamental requirement of this approach is interoperability: "the ability of two or more systems or components to exchange information and to use the information that has been exchanged."[17] Interoperability enables diverse data, tools, systems, and archives to be combined without writing custom software to handle every data link. Interoperability issues in the system-of-systems approach are accommodated through the implementation of SOA as described in *Exchanging Data for Societal Benefit: An IEOS Web Services Architecture* (USGEO 2008), which provides a suite of loosely coupled interoperable services that are available over a network and can be used across systems and platforms. Interoperability can be applied at multiple levels, including data formats and content, metadata formats and content, data access interfaces, communication networks, and power and data connections for sensor systems. To optimize interoperability, Federal agencies and relevant stakeholder communities should coordinate standards, protocols, and standard vocabulary, as well as the planning and implementing of Earth observations systems, architectures, and repositories in the framework of an integrated data-management life-cycle process. To enable participation in a true system of systems, the documentation of these baseline implementation details—data content, data dictionary, protocols, schema, and formats—must be included in the metadata.

3.2.5. Standards

Standards are common rules, conditions, guidelines, or characteristics for data and related processes, technology, and organization. The broad use of a small set of common data, metadata, and protocol standards across Federal agencies (and international standards where possible) for data development, documentation, and exchange enhances the utility of Earth observations, decreases the cost of using the data, and helps Federal agencies avoid redundancies and waste.

Different types of standards are applicable to various aspects of the data life cycle, including standards for data quality,[18] metadata standards that specify the content and structure of documentation about a data set, and interoperability standards that specify the content and structure of the digital data and how services will interact. The use of existing national and international data, metadata, and protocol standards is recommended. Because such standards are often general-purpose and require specialization for specific data types, agencies are encouraged to publish the conventions, profiles, and examples they adopt to make these standards more applicable to their data. Collaboration among and within agencies is encouraged to adopt common conventions for using existing standards to maximize compatibility.

Guidance regarding standards for different activities in the data life cycle is provided in the next section. Different communities involved in the acquisition, processing, management, and use of Earth observations have developed different standards for handling data. Through the development of the National Plan and triennial review process, these diverse standards should be harmonized where possible and new community-based standards developed where there is a convergence of interest.

3.2.6. Evaluation

Agencies should evaluate the searchability, usability, openness, and utility of their data on an ongoing basis, to determine whether they are being used for intended and unanticipated purposes; whether their use has led to positive outcomes (e.g., improved forecasts, reduced costs, and lives saved); and whether the resources required to support data distribution, use, and preservation have increased or decreased as a result of adopting a more systematic approach to data management.

3.3. Data Innovations

Agencies should actively promote innovative uses of Earth-observation data. The America COMPETES Act authorizes the use of creative techniques such as challenges and contests to spur innovations that leverage Federal resources, including data.[19] In a resource-constrained environment that demands greater returns on Earth-observing investments, data and software oriented challenges can be particularly effective mechanisms for accelerating innovation and stimulating the creation of new capabilities that bring broad benefits. Most recently, in April 2012, NASA conducted the first International Space Apps Challenge to "encourage scientists and concerned citizens…to create, build, and invent new solutions to address challenges of global importance." Such forms of targeted innovation encouragement can help to enrich the Earth-observation data ecosystem with new capabilities. Challenges can also be used to foster the development of new ideas for uses of Earth-observation data, leading to new insights regarding user requirements that can be factored into subsequent efforts.

APPENDIX A. AUTHORITIES

The National Earth Observations (NEO) Task Force was chartered in February 2011 under the NSTC CENRS. The NEO Task Force was charged with developing the foregoing Strategy, in response to Congressional direction to the Director of OSTP.[20] A December 2010 OSTP report to Congress outlined the Administration's plans for the NEO Task Force.[21] The NEO Task Force efforts build on the work of USGEO and other subcommittees of the CENRS. The following NSTC departments and agencies are represented on the NEO Task Force:

> *The Director of OSTP shall establish a mechanism to ensure greater coordination of the research, operations, and activities relating to civilian Earth observation . . .*
>
> National Aeronautics and Space Administration Authorization Act of 2010 (Pub. L. No. 111-267, October 11, 2010)

Department of Agriculture;
Department of Commerce (Co-Chair);

Department of Defense;
Department of Energy;
Department of Homeland Security;
Department of the Interior (Co-Chair);
Department of State;
Environmental Protection Agency;
National Aeronautics and Space Administration (Co-Chair);
National Science Foundation;
Office of the Director of National Intelligence;
Smithsonian Institution; and,
U.S. Agency for International Development.

The following organizations in the Executive Office of the President are also represented on the NEO Task Force:

Council on Environmental Quality;
Office of Management and Budget; and,
Office of Science and Technology Policy (Co-Chair).

The National Strategy set forth in this document outlines a process to identify, implement, and update national civil Earth observation priorities. The heart of the strategy is the development of a National Plan to guide the development, deployment, and maintenance of Earth-observing systems. As required by Congress, the plan will be updated triennially through a robust, repeatable process that ensures correspondence with changing national needs and scientific understanding. The National Plan and process will leverage advances in Earth-observing technologies both nationally and internationally and will also include external, independent advisory input.

In addition to serving the primary mission and essential functions of the U.S. Government that require Earth observations, the National Plan will support the objectives of the U.S. Global Change Research Program and serve as the foundation of the U.S. contribution to the GEOSS.[22] The plan will serve as the practical realization of the 2005 strategy for the national Integrated Earth Observation System (IEOS).[23] The plan will also support and align with relevant legislation and policies including, but not limited to, the Global Change Research Act of 1990; the Clean Air Act Amendments of 1990; the National Space Policy of 2010; the Arctic Region Policy of 2009; the National Policy for the Stewardship of the Ocean, Our Coasts, and the Great Lakes of 2010; and OMB Circular A-16, "Coordination of Geographic Information and

Related Spatial Data Activities." Both the National Strategy and the National Plan will include guidelines to support transparency and open government in accordance with the Open Government Directive[24] and the America COMPETES Reauthorization Act of 2010.[25]

APPENDIX B. REPRESENTATIVE QUESTIONS BY SOCIETAL BENEFIT AREA

The SBAs introduced in Section 1.4 are described in greater detail below. The descriptions of the SBAs are accompanied by a set of questions developed by the NEO Task Force's Policy Team to serve as a point of departure for further discussion on key Earth observation requirements. These sample questions have been culled from past agency, interagency, and National Academies' reports, as well as new input from the agencies represented on the NEO Task Force Policy Team. The questions below are notional and not prescriptive.

Agriculture and Forestry

In the agriculture sector, societal benefits accrue from measurements that can inform both short- and long-term decision making by farmers, ranchers, foresters, rangeland managers, researchers, commodity markets, and governments. By supporting the producers of renewable products, this SBA aids the ability of farmers and foresters to satisfy food, feed, fiber, inputs for biofuels, and forest product needs. Important information requirements include measurements to support (1) production decisions; (2) agricultural forecasting; and (3) greenhouse gas mitigation in the agriculture and forestry sectors; (4) verification and compliance monitoring for crop insurance; and (5) crop-production monitoring to sup-port international food security for timely famine and disaster relief. Improved data and information flow in these areas can contribute to (1) early warning systems for crop yield shortfalls and pest outbreaks; (2) understanding food, feed, forest products, and fiber needs in the context of expanding population and changing climate and biophysical resources; and (3) prevent degradation and desertification resulting from agricultural, deforestation, and grazing practices. Achieving these and other results requires effective national, regional, and local Earth observations of agriculture, rangelands, and forestry combined with comprehensive socioeconomic data at the local and regional scale. Key questions include the following:

- How can we effectively and continually monitor domestic and foreign yearly yields and harvests of food and fiber production at field, local, regional, and global scales?
- What systems are needed to develop timely feedback mechanisms at a local parcel level for within-season crop-production progress and early notification of warning factors, including pests, pathogens, water deficits, and nutrient deficiencies to enable adaptive management of agricultural production?
- How can we detect indicators of landscape health to aid in protecting crop lands, rangelands, grazing areas, and forests from degradation and desertification? How can we best measure soil erosion from wind and water to inform the understanding of tipping points for significant degradation? How can we measure land changes after natural disasters such as floods?
- In what ways are agriculture, forestry, and natural resources affected by invasive species, including plants, animals, insects, and diseases, and how can we detect and monitor the spread of disease in a timely, systematic, and synoptic manner to better derive global production from year to year? How can we track the effectiveness of remediation?
- How can we detect the effects of agro-terrorism (e.g., plant and animal diseases and waterborne pathogens), rapidly differentiate intentional infections from natural causes, and monitor progress of remediation?
- How do we monitor and manage land- and water-use competition for food and fiber production, alternative fuel sources, and urban and suburban development?
- How do we mitigate the impacts of climate change on agriculture and forestry?
- How can we detect and measure landscape factors relevant to compliance with agreements between landowners/operators and Federal and state agencies (e.g., conservations and farm programs, easements, timber sales, carbon conservation practices, and range management)? How can we monitor greenhouse gas flows from agriculture and managed forests at the parcel/field level?

- What systems are needed to detect and quantify the impact and speed of recovery from episodic catastrophic events such as drought, flood, hurricanes, tornadoes, earthquakes, and wildfires?
- How can we better monitor impacts of drought at field to county scales, to ensure equitable distribution of crop loss compensation?
- How can we effectively monitor and quantify land-use/land-cover change pertaining to agriculture and forestry, such as loss of agricultural land due to factors such as urbanization, desertification, etc.?
- How do we measure drought and characterize historical and pre-historical patterns of forest fires, including the systematic study and dating of scars in trees caused by forest fires? The pat-terns of natural wild-land fires (i.e., those that are unrelated to human activity) are a key baseline to maintaining the long-term health of natural vegetation, particularly in the southwestern United States.
- How do we effectively monitor and quantify the extent to which increases in insect, disease, and fire disturbance affect the biophysical characteristics of forests and affect the social and economic vales associated with forests?
- What is the former trends and current capacity of developed lands to sustain introduction of new crop variety or species with greater yield?
- Where do productive growth patterns appear to offer reusable elements or composites?
- How do cultural land use activities present opportunities for economic change on the margin or edge, within the network, or at the nodes?

Biodiversity

This SBA addresses genetic, species, and ecosystem-level biodiversity. Monitoring for biodiversity sup-ports strategic conservation decisions that promote balance within the ecosystems that humans depend on for their long-term well-being. This SBA provides information on the condition, extent, and diversity of ecosystems; distribution and health of species; and genetic diversity in key populations. Biodiversity measurements are important in both marine and terrestrial systems. Key questions include the following:

- What species exist, and where? Can we use remote sensing to support sustainable ecosystems enabling biodiversity preservation? How do we detect (1) native, intact communities of species of interest; (2) exotic species; and (3) regions of high species diversity?

- How do we identify the rates and geospatial characteristics of genetic variation in wild organisms? How will this help in determining spread of disease, loss of biodiversity, use of new biotechnology, and increases for global production while protecting biodiversity?
- What can biodiversity trends among widespread species tell us about how the environment around us is changing and about the sustainability of human land and ocean use at a broad scale?
- How do human activities (e.g., resource utilization, structures, disturbances, discharges, and runoff) affect biodiversity and patterns of distribution and movement of organisms?
- What can trends among the rare and threatened species tell us about localized extinction processes and about the efficacy of conservation actions at a local scale?
- What monitoring standards and sampling schemes are needed to enable non-specialists to contribute to rigorous, high-quality, population-based, time-series data on global biodiversity?
- How are human populations moving and developing, and what effects do these shifts and their accompanying infrastructure developments have on biodiversity?
- How do we understand the impacts of variable environmental forcing and climate change on biodiversity?
- What genetically and structurally unique characteristics exist that will provide new medicines, products, processes, and designs?

Climate

The climate SBA addresses the observational needs of a wide variety of climate-related activities, including modeling/projecting, mitigation, adaptation planning, and risk assessment. Climate change has wide-ranging impacts on human health, ecosystem dynamics, commerce, ocean acidification, water availability, and water quality. There is a need for long-term, consistent measurements of baseline climate data for climate monitoring, change prediction, and science, to understand critical aspects of the climate signal or fundamental Earth system processes. In addition, observation and research are needed to support regional and local adaptive-management strategies, and to inform and evaluate greenhouse-gas-mitigation policies.

Climate observations for these and other purposes can include atmospheric, oceanic, and land observations at scales ranging from a watershed to regional to global. An appropriate mix of platforms is necessary to provide needed observations at these scales. Key questions include the following:

- What is the contribution of specific observing-system capability to the confidence of climate prediction and projections, separately and collectively?
- How do trends in human systems (such as human populations and consumption changes) affect climate?
- How can we measure, track, and verify greenhouse-gas emissions reliably enough to raise the quality of "bottom-up" reporting of greenhouse gas sources and sinks?
- How will the dynamics of ecological carbon fixation, sequestration, and release change as carbon dioxide becomes more abundant?
- How do air-pollution-mitigation strategies, which may be focused on ozone and aerosols, affect global climate change, weather predictions, and air-quality concerns on regional and global scales?
- How do regional and global climate changes affect air quality?
- What changes in cloud cover, Earth's albedo, and ocean heat uptake affect the overall heat budget and warming of the planet?
- What are the probability and the likely effects of major loss of the major ice sheets? How quickly is sea level rising on average, regionally, and near low-lying population centers and sensitive ecosystems? Can we understand this better by monitoring ice-sheet-melting dynamics, mass change, and glacial hydrology?
- How do climate variations affect the intensity and frequency of smaller scale weather phenomena? To what extent do mesoscale weather phenomena and processes affect the climate and therefore need to be represented in climate models?
- How will the manner, seasonality, and regional distribution of precipitation—not just the total amount of precipitation—change with climate change? How can regional adaptation planning related to changes in water supply, farmland irrigation, and ecosystem shifts be improved through monitoring of precipitation amounts, snowpack changes, storm intensity, and other metrics?
- How can we address environmental stewardship needs in the Arctic Ocean and adjacent coastal areas in the face of climate and environmental changes?

- How does the thawing of onshore and offshore permafrost and the sublimation of coastal and offshore gas hydrates contribute to the release of CO2 and methane into the atmosphere?
- How will the rate at which the ocean stores excess heat caused by climate change be modified as elements of the climate system evolve?
- What is the extent of the urban-heat-island effect in cities, and what characteristics of cities contribute to, or could mitigate, this effect?
- How will climate change influence living marine resources and marine ecosystems, and how can we mitigate those impacts?
- How are the extremes of climate changing? Are we able to measure them, and if so, with what confidence?

Disasters

In the disasters area, societal benefits accrue from Earth observations used to monitor, predict, mitigate, respond to, assess the risk of, and provide early warning of events, including natural disasters resulting from earthquakes, landslides, droughts, coastal inundation, heat waves, tsunamis, tornadoes, winter storms, floods, wildfires, volcanic eruptions, tropical cyclones, and geomagnetic storms driven by space weather disturbances; as well as disasters that are at least partly human-induced, such as oil spills and other pollution events. The NSTC Subcommittee on Disaster Reduction has identified four key characteristics of disaster-resilient communities: (1) relevant hazards are recognized and understood; (2) communities at risk know when a hazard event is imminent; (3) individuals at risk are safe from hazards in their homes and places of work; and (4) communities experience minimum disruption to life and economy after a hazard event has passed. Changing the patterns of where we live, build businesses, and locate infrastructure could be the best option in terms of eliminating the risks communities face. Monitoring change (e.g., the Coastal Change Analysis Program) can help us understand where we are creating changes that put more people and economies at risk. All these characteristics can be achieved in whole or in part through enhanced or real-time Earth observations that embody systematic measure-ment programs to provide the capability to quantify the magnitude and impact of disasters through modeling and simulation. Key questions include the following:

- How are extreme events likely to change with changing atmospheric temperature, humidity, aerosol content, and wind patterns?

- Where are major fault systems nearing the release of stress caused by strong earthquakes, and where might associated landslides or tsunamis occur as a result? How rapidly can we characterize large earthquakes, including their tsunami potential?
- How do vulnerable populations and critical infrastructure map onto earthquake, flood, fire, landslide, and volcanic-eruption risks? Where (in real time) are important equipment and supply assets and vulnerabilities, as well as human assets?
- Where are areas with fuel loads or fuel continuity with high fire risk? How do we monitor fuels and local weather conditions at the forest level to improve our pre-fire forecasts and active fire-progression models?
- How do we detect and report new fire starts and provide active fire characterization (active fire fronts, hot spots, and dynamic fire-behavior reporting) for large fires that threaten lives, human health, ecosystems, and infrastructure?
- During preparation for major weather events, what are the likely impact, extent, timing, and intensity of events? How can we support protective measures against severe weather, such as taking shelter during tornado warnings, preparation for anticipated heavy snow, evacuation of coastal areas under hurricane threat, and rerouting aircraft to avoid severe weather?
- What are the most effective methods for sharing this information with intended audiences? How can it be most effectively coupled with existing applicable science-based information, in a format that is useful to the intended audience?
- During disaster response, what are the real-time extent, severity, and expected changes in flooding, fires, structural damage, etc.? What other potential hazards can be anticipated from leaking fuel tanks, downed power lines, washed-out roads, etc.?
- During disaster response, how can we accurately quantify the impacts and implement conservation and economic measures to minimize long-term impacts on the environment and economy?
- How can we link health databases with real-time monitoring and prospective assessment of weather, climate, geospatial, and exposure data to characterize the health impacts of extreme weather events?
- How can we better monitor and predict geomagnetic storms to mitigate the impact on electric power grids, corrosion of oil and gas

pipelines, disruption of communication and navigation satellites, and damage to spacecraft?

- How do we better image changes in the landscape—primarily topographic, visual, and thermal—that will aid in more accurate forecasts, detection, and characterization of volcanic activity? How can we more accurately track long-lived volcanic ash and aerosol clouds that affect global aviation and cause health risks in populated areas threatened by ash clouds? How can we better calibrate satellite-based measurements with ground-based observations?
- How can we use historical records of disaster occurrences within regions to better determine rebuilding requirements, change of infrastructure, need for water catchment systems above ground and underground, and crop failures?
- How do we assess impact on, and forecast recovery of, biological communities damaged or affected by disasters (e.g., hurricanes and oil spills)?

Terrestrial and Freshwater Ecosystems

In the area of terrestrial and freshwater ecosystems, societal benefits accrue from observations that enable ecological forecasting, such as the prediction of the effects of biological, chemical, physical, and human-induced pressures on ecosystems and the role of climate on ecosystems. Earth observations in this area help identify cause-effect relationships that can help managers mitigate and monitor ecosystem and climate changes, mitigate ecosystem impacts through ecosystem forecasts, and practice adaptive management. The SBA includes forests, watersheds, tundra, inland water, islands, and archipelagoes. This SBA focuses on terrestrial ecosystems. Coastal and marine ecosystems are included in the oceans and coastal resources SBA. Key questions include the following:

- Where are native and un-fragmented plant communities for grassland/shrub land/savanna, riparian, and forest cover types? Where are populations of ecologically unique endemic species and communities that are of management concern?
- What biophysical settings are necessary for the reestablishment of communities of concern? What areas have high potential to restore or reconnect endemic species populations?

- How can ecosystem-based management be facilitated through monitoring of population dynamics, migratory patterns, climate and weather forcings, and hydrology? How is the quantitative evaluation of ecosystem-nutrient dynamics affected by atmospheric dust and other aerosols?
- How is the range of the boreal (subarctic) forest likely to shift due to temperature and precipitation changes?
- How can we quantify and monitor ecosystem services at local and regional scales?
- What are the area and composition of urban and suburban landscapes, and how are these changing? What are the impacts of recent urban and suburban land-use changes on the environment, including effects on wildlife habitat, watersheds, and the water and carbon cycle?
- What are the environmental, social, and economic effects of proposed urban- or energy-development decisions under various policy scenarios and external drivers, including disasters?
- How can we monitor, authenticate, certify, monetize, and trade carbon flows from critical bio-sphere systems (e.g., global rainforests) at the regional or local level?
- In what ways can soil dynamics, including moisture, decomposition rate, and formation rate, affect large-scale patterns in vegetation and other global ecological processes?

Energy and Mineral Resources

In the area of energy and mineral resources, societal benefits accrue from observations related to fossil-fuel extraction and production; renewable energy generation (including hydropower, wind power, bioenergy, solar power, and geothermal power); energy transportation and transmission; and availability and sustainability of critical mineral resources necessary for energy technologies, economic development, defense, and future innovation. Data needs are often application-specific. For example, Landsat and ASTER radiometric data are necessary to identify minerals and alteration associated with critical resources such as rare earth elements (REE), wind speed is a critical observation set for wind turbines, insolation and cloud-cover prediction are important for solar-energy, and surface water location and volume are critical to many energy- and resource-production industries—from copper and gold, to coal and nuclear to hydropower. In all cases, a goal is to enable planners and managers to predict 20-year trends with the confidence necessary to make utility-scale investments and land use planning decisions, based on data continuity and coverage. Key questions include the following:

- What is the distribution of alteration mineralogy associated with undiscovered resources such as the strategic minerals necessary for economic development, defense, and future technology?
- How will surface-water availability change over the time scale of supporting infrastructures, facilities, and technologies established for resource extraction and fossil fuel, hydropower and nuclear power production (50+ years)?
- How do atmospheric and topographic effects commonly seen in the United States affect local wind-energy potential? Can we generate reference data sets that enable the development and validation of reliable site-scale wind models?
- What is the energy potential of offshore winds? How can we bridge the gap between ocean and coastal characterizations, since remote sensing platforms on satellites offer invaluable information on the surface winds over the ocean but have limitations in the coastal zone?
- What solar-energy resources are likely to be available within a 3-hour time frame (to enable load balancing), on a 12- to 72-hour time frame (to enable forecasting), and on a seasonal-to-interannual time frame (to enable utility planning)?
- What are the cumulative environmental impacts of large-scale resource extraction and renew-able-energy deployment in the United States? What monitoring protocols are necessary to enable regional-scale siting and adaptive management of energy and ecosystem resources (both terrestrial and oceanic)?
- What are the land-use and other environmental impacts of bioenergy programs and policies globally? How are land-use and land-cover changing around the world, and what are the bio-energy-related driving forces behind those changes?
- What are the spatial and temporal variation patterns in ocean waves, currents, and tides? Can we harness the energy they provide?
- What are the environmental impacts of mineral resource and fossil-fuel extraction, transportation, and use?
- How much energy is used in urban metabolism (the flows of the materials and energy within cities), and what are the types and

distribution of energy sources for urban and industrial energy emissions?

- How does urban and industrial energy use affect air temperature and water temperatures, aerosol loading, atmospheric chemistry, and human health?

Human Health

The health area includes societal benefits from the observation of air quality, aeroallergens, and infectious diseases, among other factors. Understanding aeroallergens and air quality depends heavily on atmospheric measurements, while infectious disease has an important waterborne component. Accelerating climate change may exacerbate a number of these stressors, including the spread of dis-ease, toxins, and contaminants. Earth observations can serve as indicators and early-warning systems for food security, water and soil quality and quantity, harmful algal blooms and seafood contamination, air quality, and other major contributors to human health. These and other indicators can be combined with socioeconomic data and infrastructural indicators to provide a rich data set related to human health and well-being. Key questions include the following:

- How does economic development in underdeveloped countries affect changes in land and water use that affect the production of land, water, and air pollutants? How are those pollutants transported on the local scale and across oceans and continents? How are those pollutants transformed during the transport process?
- How do we most effectively maintain awareness of potential changes in insect-, mammal-, and protozoan-borne disease vectors in proximity to vulnerable human populations?
- How do we most efficiently monitor air quality and air-pollution exposures and understand the emissions sources and the physical and chemical atmospheric processes that determine those exposures?
- How do we monitor the influence of climate change on, and what measurements do we need to monitor, human-health stressors such as air quality, aeroallergens, aerosolized pathogens, dust burdens, and other pollutants directly linked to asthma and other human health effects?

- What factors initiate harmful algal blooms and bacterial proliferations? How do temperature change, shifts in rainfall patterns, and other climate-associated factors affect their distribution, occurrence, and severity?
- How can data from new biosensors for pathogens, contaminants, harmful algal blooms, and toxins be connected with data from ocean and coastal observing systems and new epidemiological studies of human disease to better protect the public from ocean-borne health threats?
- How can data from observing the ocean ecosystems and biodiversity be used to develop products and biological models to enhance human well-being?

Ocean and Coastal Resources and Ecosystems

The ocean and coastal resources SBA focuses on the characteristics of, and risks to, ocean, coastal, and Great Lakes populations and ecosystems. Global, regional, and local trends in natural processes and human demands on coastal and ocean ecosystems may affect the ability of these ecosystems to support commerce, living resources, recreation, and habitation. Coastal communities are at risk from coastal hazards, development, and urbanization. The multiple stressors on coastal and ocean ecosystems, including population growth, land use, loss of biodiversity, habitat degradation, pollution, sea-level rise, and natural processes, lead to health threats to human and marine life. Observations are needed to understand interactions between coastal, marine, and terrestrial systems across the land-sea interface. Key questions include the following:

- How can we sustainably manage and monitor coastal and ocean ecosystems, particularly those that are subject to intense human harvesting?
- What air and land-based environmental contaminants are present in coastal waters?
- How is ocean chemistry changing over time (e.g., acidification due to rising CO2 or other causes)?
- How can we predict the future extent of ocean acidification and its long-term ecological con-sequences, including the capability of

marine organisms to adapt, and the impacts on sustain-ability of harvested living resources?

- How can we ensure that the U.S. ocean, coastal, and Great Lakes communities are environmentally and economically sustainable against the backdrop of natural and man-made hazards, pollution, declining ocean and coastal ecosystems, climate change, population growth, and competing uses?
- How can we ensure that goods and commerce will flow safely and effectively to and from the United States?
- How do we monitor and improve the safety of seafood in the face of natural- and human-mediated contamination?
- Which marine fish stocks are overexploited, and what are the trends in resource abundance and distribution?
- How can we strengthen the resilience of coastal communities and marine and Great Lakes environments and their abilities to adapt to climate change-impacts and ocean acidification?
- What biophysical settings are necessary for the reestablishment of communities of concern? What areas have high potential to restore or reconnect endemic species populations?
- How can monitoring population dynamics, migratory patterns, climate and weather forcings, and hydrology aid ecosystem-based management?
- How can we quantify and monitor ecosystem services at local and regional scales?
- How do human activities (e.g., resource utilization, structures, disturbances, discharges, and runoff) affect ocean and coastal ecosystems?
- What environmental factors and human activities are affecting and limiting the populations of endangered and threatened species?
- How can conservation efforts and marine resources habitat restoration be more effective or enhanced?

Space Weather

The space weather SBA focuses attention on improving understanding of the possible impacts of space weather phenomena—magnetic disturbances of the upper atmosphere and near-Earth space caused by solar activity. These currents can interfere with high-frequency (HF) radio communications, navigation signals from Global Positioning System (GPS) satellites, power grid systems, and other technological systems. These events could result in signal degradation or system loss, which could severely impact modern communication technologies. This SBA aims to improve current space weather assessment and prediction mechanisms in order to prevent and mitigate potential impacts. Key questions include the following:

- What are the potential economic and societal implications of a disruption to critical technological systems by severe space weather activity?
- What vulnerabilities do current technological systems have to space weather?
- How may future technologies be vulnerable to space weather?
- What impacts would a severe geomagnetic storm have on the U.S. electric power grid?
- What would be the immediate implications of sudden losses on the U.S. electrical grid due to a severe geomagnetic storm?
- How quickly could capabilities be restored to electrical grids in the event of a geomagnetic caused outage?
- How robust are the current capabilities to predict sudden geomagnetic storms?
- Are the current operational procedures during the event of a large storm adequate and up-to-date?
- What are the capabilities of current and planned surveillance systems?
- Currently what back-up systems are planned and implemented to support GPS signals and codes?
- How can GPS receiver technology be improved to remove ionospheric errors caused by interferences?
- What back up navigation systems are available that are independent of GPS and how do their capabilities compare to GPS?
- What would be the security, economic, and societal impacts of discrete or complete loss of satellite communication?
- What space weather measurement missions are currently planned or underway?
- What efforts can be made to gain more accurate space weather data, modeling, and space weather forecasts?
- How could space weather events impact the originally predicted lifetime of space technology systems?

Transportation

The transportation SBA encompasses the societal benefits of improving the safety and efficiency of all modes of transportation, particularly as they relate to impacts from weather events. Transportation is a primary beneficiary of improved mesoscale weather observations since the impact of weather on modes of transportation can be significant. Severe weather can compromise the safety of transit but also accumulate traffic delays which result in sustained economic loss. Additionally, damage to public transportation infrastructure could be mitigated through improved understanding and prediction of weather events. Key questions include the following:

- How can weather-related transportation crashes be reduced?
- How can weather-induced delays be mitigated to minimize economic loss?
- What pre-existing technologies can be employed to improve safety in transportation during severe weather?
- How far reaching are the economic impacts of weather-related delays in transportation?
- How can infrastructure damage caused by weather be reduced to save money and protect public property?
- How can delays to air-traffic due to weather be minimized?
- How can real-time observational weather data be applied to improve transportation safety?
- How can the environmental impacts and costs associated with chemical anti-icing and de-icing materials be minimized?
- Since track washouts can severely impact safety and cause large amounts of damage, how can this type of damage be prevented?
- How can port operations be optimized following weather damage?
- What improvements can be made to maintain and improve roadside meteorological observing stations?
- How can roadside data be utilized in real-time safety and traffic decisions?
- How can the reliability of individual state data be made less variable?
- How can sensor data be made more readily accessible by neighboring states, decision makers, and transportation consumers?
- What developments can be made to advance integrated observational systems and data management for surface transportation?

- How can observational data be individualized for the observational needs of each relevant mode of transportation (i.e., railroad needs versus air transport)?
- How can we most efficiently route air-traffic to avoid volcanic ash and sulfur dioxide?
- How can we optimize routing efficiency for both air and maritime transportation?
- How can we best route ships though ice fields?

Water Resources

Societal benefits accrue from observations related to water-resource management, emergency management, tourism, and recreation. This area focuses on terrestrial hydrology, including surface water, groundwater, terrestrial hydrological forcings, water quality, and water use. Oceans and coastal margins are covered in the oceans and coastal resources SBA. Improved understanding and management of the water cycle has wide-ranging implications for food security, energy development, human health protection, and critical habitat protection. Key questions include the following:

- What is the sustainable yield of a given watershed? How have the characteristics and health of the watershed changed, on seasonal to decadal scales?
- Where will changes in water supply affect the viability of nuclear, hydroelectric, or innovative power generation?
- How are changing snowpack, rainfall, surface water, and groundwater affecting water-storage needs in the United States?
- How have water-usage patterns changed over time? How have those changes affected water quality and supplies?
- How do precipitation, temperature, evaporation, and other hydrologic processes influence water availability and quality?
- What are the transport and fate characteristics of chemicals, nutrients, sediments, pathogens, harmful algal blooms, toxins, and marine debris in waterways?
- To protect drinking-water intakes and reduce unnecessary and expensive beach closures, what systems and networks are needed to advance the rapid detection of contaminant plumes and pathogens in rivers and recreational waters?

- How are changes in evaporation, precipitation, soil moisture, and human water usage affecting the balance of the water cycle on local to regional scales?
- How can we better monitor water use by individuals in areas where there are water-rights conflicts, and how can we supply the prior and current consumption data required to negotiate interstate and international water compacts?
- How can we identify regions or countries where freshwater resources are becoming scarce, posing a potential threat to sociopolitical stability and food security?

Weather

Societal benefits accrue from observations that support improved weather information, forecasting, and warning, including numerical weather prediction, nowcasting, and atmospheric-chemistry monitoring. Social and economic sectors ranging from agriculture to aviation are directly affected by changes in temperature, precipitation, winds, and other general weather properties. Improvement in data sets and predictive capacity can enable these and other industries to plan for and respond to weather with more confidence. The weather SBA cuts across several other SBAs. Long-term observations of weather changes are related to climate observations, and low-frequency/high-impact events are closely tied to the disasters SBA. Weather prediction to support renewable energy development is also connected to the energy SBA. Key questions include the following:

- How do improved weather forecasts affect regional economic decisions such as pre-positioning store items in anticipation of increased demands (e.g., snow shovels in advance of heavy winter storms, water and canned goods in preparation for tropical storms, or umbrellas during periods of heavy rain).
- How do improved weather forecasts support emergency managers in preparation for, or response to, developing weather conditions? For example, emergency managers may delay or cancel public events, call in additional snow plow operators, or lift high-occupancy-vehicle controls, in response to weather information tailored to their requirements.

- How do meteorological effects, including convection, turbulence, and clouds, affect transport of chemical compounds and atmospheric chemistry?
- In what ways do weather and climate services need to be tailored to urban environments? How can we provide the level of resolution and surface specificity necessary to produce forecasts on the urban scale?
- How can we improve mesoscale predictive capabilities (tens to hundreds of kilometers) to tailor weather impacts to specific local areas and increase the effectiveness of preparation?
- How do convection, chemical transport, and winter precipitation types evolve throughout the lowest 2 to 3 kilometers of the atmosphere?
- In what ways do precipitation patterns vary on the mesoscale? Can understanding this variability enable quantitative precipitation forecasts to enable more effective planning and response to weather events?
- How can we reduce loss of life and property, and minimize disruption from high-impact weather events?
- How do soil moisture and evaporation patterns on the landscape affect drought persistence (precipitation recycling)?

Reference Measurements

Internationally recognized references and measurement standards are the foundation upon which all Earth-observation data and analyses rest. From these are derived the range of fundamental reference measurements critical to Earth observations. These include, but are not limited to, geodesy (measurement of Earth's shape and gravitational field), bathymetry (the study of the depth of oceans and lakes), topography (the measurement of land elevation), and geolocation (precise time and global positioning). These reference measurements, among others, are the foundation on which all Earth-observation data and analysis rest. It is critical to evaluate the requirements and coverage of reference measurements in any comprehensive prioritization of Earth observation needs, taking into account ground-truths when possible. In addition, overlap between congruent satellite missions should be preserved wherever possible, to ensure inter-calibration and long-term continuity/comparability of record.

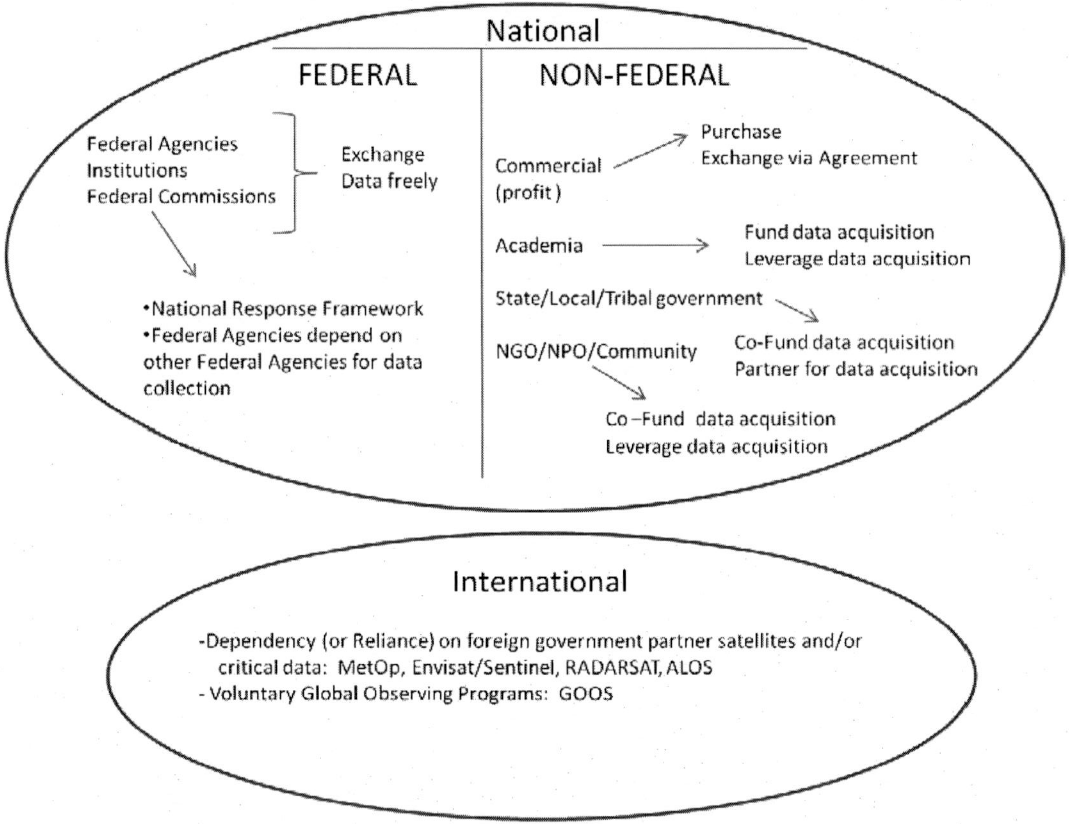

Figure 3. Relationships between Federal, non-Federal, and international stakeholders.

APPENDIX C. INTERDEPENDENCIES AND PARTNERSHIPS

The Nation's Earth-observing requirements are met through a combination of systems, including those that are entirely owned and operated by the Federal Government as well as "extramural" systems sustained by a complex patchwork of funding arrangements from both Federal and non-Federal sources. In many cases, Federal mission-driven agencies have become reliant on data from extramural systems for their operations. Often these are long-term, time-series data needed to support scientific understanding of climate change. Funding streams for extramural Earth observation systems are often interconnected, requiring the perpetual agreement of all parties that funding not be arbitrarily withdrawn. Despite best efforts, the stability of even the most important of these systems is routinely challenged by threats to their partial funding arrangements, from both Federal and non-Federal partners. While greater fiscal stability is needed in this domain, Federal agencies and partner entities have also worked to achieve substantial cost savings. USGEO is committed to tracking such efforts and highlighting the best arrangements to apply best practices and lessons learned.

APPENDIX D. DATA PRINCIPLES AND DEFINITIONS

D.1. Data Principles

- **Full and Open Access** This National Strategy adheres to the principle that access to data-managed or paid for using Federal funds should be open to the public as soon as possible after collection, in a nondiscriminatory manner, and generally free of charge. This National Strategy respects the "presumption of openness" called for in 2009 Open Government Directive. Full, open, and timely sharing and exchange of data, products, and services will through machine-readable formats and APIs (Application Programing Interfaces) enable users to effectively and rapidly transform data into useful information for society. The timeliness, accessibility, and cost may not be the same in all cases (e.g., for routine, ongoing observations by automated sensors versus sporadic, labor-intensive data collection) but exceptions to this principle should be rare and explicitly justified on a case-by-case basis. For example, access to certain fishery-dependent data (observers, location of catch, and effort) are subject to confidentiality requirements that may affect the circumstances of their release.

- **Timeliness** Data should be made publicly available with minimum time delay after capture. Data processing, quality control, and assurance processes should be automated, whenever possible, to minimize delays. In limited circumstances, some scientific investigations may permit a temporary data hold before distribution.
- **Non-discrimination** Data should be made publicly available in a platform-agnostic manner to the widest community possible. Earth observations should be distributed in a way that does not unfairly hinder access to that data on any basis unless a specific exemption has been granted. Possible exceptions to open access include data whose public dissemination is prohibited by law (e.g., personally identifiable or proprietary information), international agreements, or for reasons of national security (e.g., classified information).
- **Minimum Cost** Data should be made available free of charge to the greatest extent possible. Whenever possible, data should be made available and accessible online via web services or other Internet-based mechanisms. Agencies may set user charges in a manner consistent with OMB Circular A-130.

These principles are exemplified by NASA's Earth Science data policy, which was almost universally "free-of-charge" even before the majority of NASA's data was available and accessible online. This has been long recognized at NASA as a key factor in the wide use of these data by stakeholder communities. Additionally, some programs, notably Landsat, were originally required to charge for data in an effort to recoup operations costs. Since USGS made the Landsat archives available free of charge over the Internet, public use of the data has dramatically increased.

- **Preservation**
  Earth observations are not reproducible after the moment of measurement has passed, and are often acquired using costly technologies, including satellites, advanced sensors, and open-ocean buoys. Therefore, these observations should be managed as agency and national assets. This means they should be preserved for future use and protected from unintended or malicious modification. The data should not merely be preserved as-is but should be made "durable." Some data deemed not to have long-term usefulness may not merit preservation; however, the decision not to preserve should

be a conscious one rather than an accidental failure to preserve. NOAA's "Procedure for Scientific Records Appraisal and Archive Approval: Guide for Data Managers"[26] provides an example of agency-level guidance in this appendix.

- **Information Quality**
  Earth-observation data and ancillary information should be of known quality, meaning that they are fit for their intended use in operational missions and for planning. There may be times, however, when data must be distributed before quality control and documentation have been completed. For example, during the Deepwater Horizon oil spill in the Gulf of Mexico the responding agencies at times needed to share raw data immediately among personnel who had the expert knowledge to recognize and account for the limitations associated with those data. When data and web services are provided on an as-needed basis, quality parameters and uncertainties should be explained and clearly communicated to users. Quality control is discussed in more detail as part of Appendix D.3.
- **Ease of Use**
  To encourage the broadest possible use of Earth observations, users should be able to find the data easily. The data should be made available and accessible online via web services or other Internet-based mechanisms, rather than by shipping physical media or by establishing dedicated or proprietary linkages. These services should comply with non-proprietary, interoperability specifications for geospatial data. Where possible, Data should be offered in formats that are known to work with a broad range of scientific or decision-support tools. Common vocabularies, semantics, and data models should be employed. Feedback from users should be gathered and used to develop usability improvements. Users should be able to unambiguously cite datasets, both for later reuse and to provide credit and traceability to the originator.

D.2. Metadata

Metadata provide information about data provenance, descriptive attributes, quality, accuracy, maturity, or the underlying structure in which data are stored. Metadata support interoperability with other systems, archives, and databases to facilitate data discovery and sharing, and are critical for documenting and preserving agencies' data assets. Reliable metadata, structured in a standardized manner, are essential to ensuring that data are used appropriately, and that any resulting analysis is credible.

The core metadata standards for national data are the FGDC Content Standard for Digital Geospatial Metadata and the International Organization for Standardization (ISO) standards in ISO 19115 (content) and ISO/TS 19139 (XML schema).

Certain types of metadata, such as a field-level dictionaries, documentation of units of measure, or coordinate reference system coordinates, are indispensable to any use or reuse of data and, therefore, must be collected and supported by discovery. Efforts should focus on creating a common baseline for metadata to achieve increased interoperability while enabling agencies to build more specific requirements on top of the baseline for individual data sets. Further, access must be incorporated into protocols to preserve the utility of the data. Additional descriptive metadata may include information that helps contextualize data, such as descriptions of measurements or analytic techniques, and are essential to their reuse beyond their original purpose. Inclusion of a data dictionary or schema as part of data metadata will help assure users to evaluate the data's fitness for use for a specific application domain. Metadata for data, products, and web services should be collected and provided using international standards to promote consistency and interoperability, support discovery through catalog search mechanisms, and support fitness-for-use assessments by potential end-users. Metadata should include as much information as possible regarding not only quality but source, lineage, responsible parties, and known limitations of the data.

D.3. Data Lifecycle Activities

Planning and Production Activities

Planning and production activities in the data life cycle comprise: requirements definition, planning, development, deployment, and operations. These include such tasks as assessing the need and requirements for a new observing system, planning how to meet those requirements, developing any necessary sensors, deploying the observing system, and operating and maintaining the observing system. We do not cover these topics in this annex, focusing instead on the management of actual data once observations are collected. Nevertheless, activities that occur later in the data life cycle may influence this phase. For example, a calibration error discovered through quality-control processes may lead to changes in the operating procedure.

Data-Management Activities

Data-management and usage activities in the data life cycle include:

- **Data Collection and Processing**
  These activities refer to the initial steps of receiving data from an Earth-observing sensor (or running a numerical model that generates data), ingesting the data into the initial storage location (which may not be the final archival site), and performing the processing necessary to transform raw data from the sensor (which may include telemetry or other ancillary information) into usable data records.
- **Quality Control**
  Data quality standards are highly dependent on the nature of the data and the purpose for which they are used. Agencies should clearly and publicly document their data standards and the process by which those standards are applied to assess the quality of actual data. Agency data should be of known quality, which means that the result of the quality-control process, and a link to descriptions of the process and standard, are available in the data documentation.
  Quality-control (QC) tests should be applied to the data, including both automated QC in near-real-time and delayed-mode, human-assisted checks. Quality-assurance (QA) processes should be applied to provide validation that Earth-observation data meet, and continue to meet, their intended requirements throughout the Data Lifecycle. All QC and QA checks should be publicly described and peer-reviewed. The results of these checks should be included in metadata. All bad, suspect, and missing values should be flagged as such.
  Information quality is composed of three elements—utility, integrity, and objectivity (U.S. Department of Commerce, Section 515 Standards). Data quality will be ensured and established at levels appropriate to the nature and timeliness of the information to be disseminated. Utility means that disseminated information is useful to its intended users. Data integrity refers to security—information must be safeguarded from improper access, modification, or destruction, to a degree commensurate with the probability and magnitude of harm that could result from the loss, misuse, unauthorized access, or modification of such information. Objectivity ensures that information

is accurate, reliable, and unbiased, and that information products are presented in an accurate, clear, complete, and unbiased manner.

This framework, wherever possible, should follow the guidelines of the Data Quality Act, which directs the OMB to issue guidelines that "provide policy and procedural guidance to Federal Agencies for ensuring and maximizing the quality, objectivity, utility, and integrity of information (including statistical information) disseminated by Federal Agencies." These guidelines are designed so that agencies can incorporate the standards and procedures required by these guidelines into their existing information resources management and administrative practices rather than create new and potentially duplicative or contradictory processes. Agencies need only ensure that their own guidelines are consistent with OMB guidelines, and then ensure that their administrative mechanisms satisfy the standards and procedural requirements in the new agency guidelines. This assurance is a pre-requisite for agency postings to Data.gov, now a primary catalog for all government data, including Earth-observation data. Furthermore, this framework should follow the principles of the intergovernmental GEO "Quality Assurance Framework for Earth Observation" (QA4EO), which includes associated QA/QC information with data and datasets to support harmonization and interoperability, and enable reproducibility.

- **Documentation**

  Documentation activities produce metadata as defined in Annex D.2. Documentation may occur multiple times during the data lifecycle, and may be performed by a variety of systems and individuals. For example, basic information about the sensor system, location and time are available at the moment of data collection; subsequent quality control provides additional metadata; later analysis and use of the data may result in new or revised documentation. It is important that these documentation steps be performed at the appropriate times by qualified personnel, recorded in a standard fashion, and bundled or associated with the data.

- **Dissemination**

  Data should be made available and accessible online via web services or other Internet-based mechanisms, rather than by shipping physical media or by establishing dedicated or proprietary linkages. Immediate access to online data is preferable to delayed access via ordering services to near-line or off-line data. This goal is not always achievable for high-volume data collections, in which case agencies should carefully consider and optimize caching algorithms based on usage tracking to maximize the likelihood of popular data being online.

Dissemination may be accomplished in a "pull" mode, whereby users request data they desire on an ad hoc or periodic basis, or in a "push" mode whereby the data provider sends data to recipients on a subscriptions basis.

Agencies should consider distributed virtual storage in the Internet "cloud," especially for popular datasets, which can be helpful in reducing storage and bandwidth requirements in their data centers. See the U.S. Chief Information Officer's 25 Point Implementation Plan to Reform Federal Information Technology Management (2010) for more on this topic.

Data should be offered in formats that are known to work with a broad range of scientific or decision-support tools. Common vocabularies, semantics, and data models should be employed.

- **Cataloging**

  "Cataloging" is used here in a general sense to refer to all mechanisms established by data providers through which users are able to find the data they need. The word "discovery" is employed to refer to the user's act of finding data. Cataloging thus enables discovery.

  Federal agencies should seek to ensure that Earth observations are readily accessible to the diverse community of domestic and international information providers and users. Data collected for a particular purpose can only be reused for novel purposes if those unanticipated users are able to discover the data. Furthermore, as science progresses and policy debates sharpen, the discussion of ideas and dissemination of findings depends critically on the ability to find relevant data outside of one's agency or normal information community. For example, it is often useful to combine data from Earth observation sensors with socio-economic analyses or political and cultural assessments to perform impact and options studies.

  Cataloging methods include establishing formal standards-based catalog services, building thematic or agency-specific portals, enabling commercial search engines to index data holdings, and implementing emerging techniques such as feeds, self-advertising data, and casting. Catalog Services are the current dominant mechanism for the discovery of geospatial data and information sources and have been crucial for those who seek to understand or apply Earth-observation data and information. Accordingly, interoperable cataloging technologies for information resources and data holdings comprise a foundational requirement of the Data-Management Framework.

- **Preservation and Stewardship**

  Long-term data durability requires both preservation (ensuring the data are stored and protected from loss) and stewardship (ensuring the data continue to be accessible, e.g., by migrating to new storage technologies) and are updated, annotated, or replaced when there are changes or corrections.

  A major challenge to long-term preservation of data is the ability of the hosting agency to guarantee the authenticity and quality of its digital holdings over time. Therefore, data archives will include easily accessible information about data holdings, data structures, and data products and will include quality assessments and supporting information for efficient data discovery and retrieval. Federal Earth observation archives will adhere to data archiving guidelines that meet or exceed current national and international standards.

- **Usage Tracking**

  Usage tracking refers to the ability of the agency to measure whether their data are actually being used. Crude estimates can be made by counting data requests or data-transmission volumes from Internet servers. Such statistics, however, do not reveal (1) whether data that were obtained were actually used; (2) if the data were used, whether they were helpful or had an impact; or (3) whether the initial recipient redistributed the data to other users.

  More sophisticated means of assessing usage—while preserving the anonymity of users—are desirable. To enable better usage tracking, data should be made available through application programming interfaces (APIs); it may be possible to capture more insightful information about how data and web services are being used.

- **Final Disposition**

  Agencies should establish a procedure to determine the final disposition of data. Not all data and derived products must be archived. Agencies should establish a process whereby an evaluation regarding what to archive is performed for each class of Earth observation. Ideally, because an observation cannot be repeated once the moment has passed, all observations should be retained, but tradeoffs are possible. For example, perhaps derived products that most users have access to may adequately replace raw data and processing algorithms. Model-generated data, which are often quite voluminous, similarly pose the question of whether to archive the model outputs or merely the model code and initialization. Software algorithms and models do not require much storage space, but ensuring the ability to run the code in perpetuity may be difficult.

Usage Activities

Activities related to data usage in the data lifecycle comprise:

- **Discovery and Reception**
  Users must be able to find and access the data they want. These activities are enabled by the dissemination, cataloging, and documentation actions discussed above.
- **Analysis**
  "Analysis" includes such activities as a quick evaluation to assess the usefulness of a data set, including a data set among the factors leading to a decision, or an actual scientific analysis of the data in a research context. Such activities are only possible if the data have been well-documented and are of known quality.
- **Product Generation**
  Some uses of data involve creating new data products, such as by averaging, combining, differencing, interpolating, or assimilating data. These new products may themselves constitute a new source of data that merits its own life-cycle data-management process. Agencies are encouraged to relate the products they generate back to the original source data via appropriate documentation and catalog cross-referencing.
- **User Feedback**
  The design, implementation, and use of Earth-observing data systems and processes should take usability, defined by ISO as "the extent to which a product can be used by specified users to achieve specified goals with effectiveness, efficiency, and satisfaction in a specified context of use," into account as part of an integrated data-management life-cycle process. Data users should have a mechanism to provide feedback to the agency regarding ease of use, suspected quality issues, and other aspects of the data. Concerns should be evaluated through

the lens of community standards by Federal agencies and by the relevant user groups that rely on Earth-observation data.

- **Citation**
  Citation refers to the ability to unambiguously reference a data set that was used as input to a model, a decision, or a scientific paper. This is a topic of ongoing interest that has not been solved. Workshops in 2011 included Geo-Data Informatics: Exploring the Life Cycle, Citation and Integration of Geo-Data[27] sponsored by the NSF and Developing Data Attribution and Citation Practices and Standards[28] sponsored by the National Academy of Sciences (NAS) Board on Research Data and Information (BRDI).
- **Tagging**
  Tagging refers to the ability to identify a data set as relevant to some event, phenomenon, purpose, program, or agency without needing to modify the original metadata. The ability to tag is essential because the current practices of (1) creating new catalogs and asking people to re-register a relevant subset of their data there or (2) asking people to add new metadata tags such that an external project can detect them (cf. GEOSS DataCORE) are not scalable because they require additional work and lead to the proliferation of duplicate data sets and metadata records. Existing examples of tagging today include the ability of Facebook users to tag individuals in a photo or of desktop photo-management software to tag all photos of "Italy" or "sunsets" without modifying either the file-level metadata or the folder hierarchy. Tagging relates to cataloging in the sense that tags are additional metadata associated with another metadata record.
- **Gap Analysis**
  Gap analysis refers to the determination by users that more data are needed. Such a determination influences the requirements-definition activity that is the start of a new data life cycle.

Some aspects of these activities are outside the scope of agency data management *per se*. Once users have obtained the data they seek, agencies may have little or no knowledge or control over how those data are used. Nevertheless, the ability to obtain and use data is a primary indicator of a good life-cycle data-management process, and information from or about users and their data-seeking experience can be a valuable tool as agencies seek to improve their data management processes.

ANNEX A. NATIONAL EARTH OBSERVATIONS TASK FORCE FORMATION

Strengthening Federal Mechanisms for the Coordination of Civilian Earth Observations *December 22, 2010*

I. The Value of Integrated Earth Observations

Each day, the scientific community collects millions of individual Earth observations, allowing us to examine, monitor, and model atmospheric composition, seismic activity, ecosystem health, weather patterns, and hundreds of other characteristics of our planet. Through these national investments in Earth observations, the U.S. government ensures that our Nation's decision makers, businesses, farmers, health care workers, and indeed all our citizens have the information they need to cope with weather-related and other environmental threats to human well-being.

The myriad of observations taken today vary widely in purpose and scope and are distributed among hundreds of programs under the purview of Federal agencies and other institutions and individuals. The Obama Administration recognizes that a coordinated approach is needed to cost-effectively sustain and build on the current set of Earth observations. Beyond coordination, there is a need to ensure that Earth observations are maintained and continued with diligence commensurate with their economic and environmental importance.

II. Congressional Direction

Section 702 of the NASA Authorization Act of 2010, signed into law on October 11, 2010, provides that the Director of the Office of Science and Technology Policy (OSTP) shall establish a mechanism to ensure greater coordination of civilian Earth observations. This should include the development of a strategic implementation plan that is updated at least every 3 years. The Director shall provide a report to Congress within 90 days of enactment on the implementation plan for this mechanism.

III. Planned OSTP Actions in Response to Congressional Direction

In proceeding to strengthen civilian coordination of Earth observations, OSTP, in collaboration with other offices within the Executive Office of the President (EOP) and federal agency representatives, will establish a National Task Force on Earth Observations to draft a National Strategy for Earth Observations. Among other elements, the National Strategy will consider:

a) *A review of issues and options concerning the challenge of transferring civilian Earth observation measurement responsibilities from a demonstration agency to a long-term data acquisition agency.* The strategy will examine critical decision points where research missions should be evaluated, possible criteria for determining whether a particular observational capability should be made operational, and recommendations for facilitating the transfer of meritorious research missions to operational use.

b) *Criteria for prioritizing among observing systems as part of the budget process.* The strategy will approach how best to measure Earth system parameters with a view toward optimizing and prioritizing the mix of satellite and *in situ* systems. In addition, the strategy will consider the nation's investment in satellite operations, data-management, and information-delivery systems as an integral part of the national observing system infrastructure. The strategy will build on the OSTP report released to Congress in September 2010, *Achieving and Sustaining Earth Observations: A Preliminary Plan Based on a Strategic Assessment by the U.S. Group on Earth Observations.* Once completed, the strategy will serve as standing guidance for preparation of agency Earth observation programming and budgets over time.

c) *A strengthened national governance and coordination mechanism to support the implementation of the strategy and address its recommendations.* The governance mechanism will evolve from the existing work and coordination efforts of the United States Group on Earth Observations (USGEO), a subcommittee of the National Science and Technology Council's Committee on Environment Natural Resources and Sustainability (CENRS). While acknowledging agency mission priorities and working with existing channels, the mechanism will be designed such that agencies are responsive to its guidance in the formulation of their budgets, and through it, the Office of Management and Budget (OMB) can enhance cross-agency integration of Earth observation budget priorities. The governance mechanism will also function as the national coordination point for interaction with the intergovernmental Group on Earth Observations (GEO) and other international activities and agreements relating to Earth observations. Further, the mechanism will draw upon the cross-agency efforts to implement the Open Government Directive and enhance Data.gov with more high-value data and data services.

The task force will be co-led by OSTP, the National Aeronautics and Space Administration (NASA), the National Oceanic and Atmospheric Administration (NOAA) of the Department of Commerce, and the United States Geological Survey (USGS) within the Department of Interior. Its membership will include the organizational, member-agency principals of the CENRS and senior officials from within the EOP appointed by the Assistant to the President for Science and Technology.

The task force will be established by February 2011 and complete its interim recommendations by May 2011. Following a full CENRS concurrence, the task force will submit its interim recommendations to the National Academies for a review and comment period not to exceed three months. The task force will revise the strategy based on comments from the National Academies, and the Administration will release the National Earth Observation Strategy based on those revisions.

IV. Approach to a National Strategy for Earth Observations

Since early 2010, OSTP has consulted widely across Federal agencies to gather viewpoints on the best approach for developing a national strategy for Earth observation. From these consultations, OSTP concludes that:

1. Different agencies are usually responsible for research and operations, creating discontinuities in the way Earth observation assets (satellite, sensor, and measured) are developed, managed, operated, and funded. A sustained commitment to coordinated satellite system and sensor development is necessary to make effective use of taxpayer investments and avoid a loss of observing capability in the next decade.
2. Coordinated attention to *in situ* systems is needed.
3. The strategy must examine the process for appropriately prioritizing investments in the development of Earth observation applications, so that the nation can fully capitalize on its major Earth observation investments. The strategy must also allow for a critical level of research and experimentation to complement those investments made to fulfill user-driven needs.
4. The strategy must consider and address the observational needs of the US Global Change Research Program (USGCRP), the National Climate Assessment (NCA), the Interagency Climate Change Adaptation Task Force, the Climate Change Technology Program (CCTP), the Climate Roundtable on Climate Information and Services

(RCIS), the National Ocean Council (NOC), and the Administration's other clean energy, environmental, and climate change mitigation efforts.

5. The strategy must support and build on the appropriate language and concepts in the National Space Policy, the Arctic Region Policy, and the National Policy for the Stewardship of the Ocean, Our Coasts, and the Great Lakes. The national strategy must also take account of the previous work of the USGEO strategic assessment, delivered in the OSTP September 2010 report to Congress on Earth observations. In addition, the national strategy should take account of the work of the Office of the Federal Coordinator for Meteorology (OFCM), as well as existing plans for Arctic observations, the U.S. Integrated Ocean Observing Systems (IOOS), agricultural, and other systems.
6. The strategy and governance mechanisms must support transparency and open government, the Open Government Directive, and leverage the evolving Data.gov infrastructure for delivery of data and data services.

V. Conclusion

One of the primary functions of OSTP is to provide the leadership and needed coordination of national Earth observation activities. In assembling the National Task Force on Earth Observations, OSTP will bring together expertise from across the government, drawing from each agency's distinctive capacity, to construct the relationships and interactions among the agencies that will result in a program for Earth observations that contributes to both our national prosperity and our national security.

VI. Additional Background: GEO, USGEO, and the Visionary Frameworks of GEOSS and IEOS

The concept of integrated Earth observations achieved international prominence in 2005 with the establishment of the intergovernmental GEO and its agreement to build a Global Earth Observation System of Systems (GEOSS) over 10 years. This concept has been endorsed and is currently being implemented by 85 governments, the European Commission, and 58 international organizations. The United States co-leads this international activity with China, the European Commission, and South Africa. GEOSS represents a grand vision for comprehensively observing the Earth system and integrating the data gathered from such observations into useful information for society.

Concurrent with the establishment of GEO, OSTP formed a working group on Earth observations under the National Science and Technology Council (NSTC), which published its *Strategic Plan for the U.S. Integrated Earth Observation System (IEOS)* in late 2005 as the U.S. national contribution to GEOSS. This working group was later re-constituted as USGEO, and is now a standing subcommittee of the CENRS. USGEO was chartered to begin implementing the U.S. IEOS and to serve as the national coordination mechanism for making input to GEOSS.

Since 2005, USGEO has released a number of important reports that clearly articulate those areas of opportunity that are ripe for national investment in Earth observation, and maintained a substantial U.S. engagement with the intergovernmental GEO. Most recently, USGEO conducted a strategic analysis of the nation's Earth observing requirements, resulting in a valuable list of 17 Earth system parameters and corresponding observing systems considered most essential for achieving the societal benefit areas identified in the IEOS plan. This analysis was the foundation of the OSTP report released to Congress in September 2010, as a first step toward a national strategy for Earth observations.

ANNEX B. HISTORY OF UNITED STATES GLOBAL CHANGE RESEARCH PROGRAM DATA PRINCIPLES

The data requirements for understanding global change were tabulated intensively by NASA Headquarters during the mid- to late-1980s, working with the broad Earth observation user community within and outside the U.S. Government. An early team called the Earth Observation System (EOS) Science and Mission Requirements' Working Group defined a suite of instruments for acquiring measurements of the Earth's atmosphere, surface, and interior from low Earth orbit; an information system to support scientific research; and a vigorous program of scientific research, stressing study of global-scale processes that shape and influence the Earth as a system. As a result, NASA's Earth Observation System Program was initiated in 1991.

At that time, NASA established a policy on the availability to the public of NASA-acquired Earth science and applications data. A major impetus of this new policy was the tenet that data would be made avail-able to all users as soon as practicable after acquisition and without any period of exclusive access for any user group. The awareness of the importance of an open data policy grew out of unsatisfactory policies from the 1980s which allowed periods of exclusive use for certain user groups on key data sets.

Under the policy as originally conceived, the data would be provided to research users at a price not to exceed the marginal cost of reproduction and distribution. Later, in 1994, after consultation with the user community and study of current practices for NASA missions and at NOAA and USGS, NASA Headquarters waived this marginal cost of reproduction and distribution. NASA's aim was to promote the widest possible use of data, with a science-/user-driven system. The stated goal was not only to advance the science of remote sensing but to increase understanding of the Earth system. All research and applications demonstration users were asked to provide results for publication in scientific literature or as technical reports. Government-funded investigators were expected to make available research results, models, algorithms, and other supporting information to permit replication of research, consistent with U.S. law and policy.[29]

Data Management for Global Change Research Policy Statements

July 1991

The overall purpose of these policy statements is to facilitate full and open access to quality data for global change research. They were prepared in consonance with the goal of the U.S. Global Change Research Program and represent the U.S. government's position on access to global change research data.

- The Global Change Research Program requires an early and continuing commitment to the establishment, maintenance, validation, description, accessibility, and distribution of high-quality, long-term data sets.
- Full and open sharing of the full suite of global data sets for all global change researchers is a fundamental objective.
- Preservation of all data needed for long-term global change research is required. For each and every global change data parameter, there should be at least one explicitly designated archive.

Procedures and criteria for setting priorities for data acquisition, retention, and purging should be developed by participating agencies both nationally and internationally. A clearinghouse process should be established to prevent the purging and loss of important data sets.

- Data archives must include easily accessible information about the data holdings, including quality assessments, supporting ancillary information, and guidance and aids for locating and obtaining the data.
- National and international standards should be used to the greatest extent possible for media and for processing and communication of global data sets.
- Data should be provided at the lowest possible cost to global change researchers in the interest of full and open access to data. This cost should, as a first principle, be no more than the marginal cost of filling a specific user request. Agencies should act to streamline administrative arrangements for exchanging data among researchers.
- For those programs in which selected principal investigators have initial periods of exclusive data use, data should be made openly available as soon as they become widely useful. In each case, the funding agency should explicitly define the duration of any exclusive use period.

Source: Data Management for Global Change Research Policy Statements, U.S. Global Change Research Program, July 1991, www.gcrio.org/USGCRP/DataPolicy.html.

Congress mandated the United States Global Change Research Program in the Global Change Research Act of 1990 (Public Law 101-606). OSTP organized and worked with an interagency team composed of agency members of the USGCRP to create the data principles now commonly known as the "Bromley Principles," named for D. Allen Bromley, who was Science Advisor to President George H. W. Bush at the time. A statement of national policy, issued in July 1991, applied to all U.S. Government agencies, and established the framework within which agencies should work:

With these principles in mind, the United States also drove advances in international data policy in the early 1990s. NASA organized a group of partners to participate in the "International Earth Observing System," known as IEOS. The partners, including NASA, European Space Agency (ESA), European Organisation for the Exploitation of Meteorological Satellites (EUMETSAT), NOAA, Canadian Space Agency (CSA), National Association of State Departments of Agriculture (NASDA), and the Japanese Ministry of International Trade and Industry (MITI), were to work together to establish international data exchange principles for:

- NASA Earth Observing System
- ESA Polar Orbit Earth Observation Mission (POEM)
- NOAA POES series after NOAA-N
- NASA and NASDA's joint TRMM mission
- NASDA Advanced Earth Observation Satellite (ADEOS)

The U.S. civil space agencies NASA, NOAA, and USGS then used the Bromley Principles as background in the international Committee on Earth Observation Satellites (CEOS)[30] to develop an agreement by all CEOS members on Data Exchange Principles. These data exchange principles were meant to eventually cover the full range of remote sensing missions, focusing first on data exchange principles in support of global change research, and then continuing development of CEOS data principles beyond research use. At the Sixth CEOS Plenary meeting held in London in December 1992, the Resolution on Satellite Data Exchange Principles of Support of Global Change Research was adopted. These principles were developed at an ad hoc CEOS Data Policy Meeting hosted by CNES in Paris in October 1991, and they elaborated on the principles first adopted in 1991. The eventual CEOS Data Principles in Support of Operational Environmental Use for the Public Benefit were developed at an ad hoc Data Policy Meeting hosted by NOAA and NASA in Washington, DC, in April 1994. The principles were adopted at the eighth CEOS Plenary meeting held in Berlin in September 1994 and first published in the 1995 CEOS Yearbook.[31]

The first Earth Observation Summit convened in Washington, DC, in July 2003 and was attended by high-level officials from 33 countries, the European Commission, and 21 international organizations involved in Earth observations. The *Global Earth Observation System of Systems (GEOSS) 10-Year Implementation Plan*, published in February 2005, contains the GEOSS data-sharing principles (Section 6.1 Key principles).

These valuable and enduring U.S. data policies for Earth observation, and the concomitant exercises with our international partners, arose from widespread recognition of the importance of research and monitoring for global environmental understanding and sharing of data to support the public good. The observing capability to achieve these shared goals exceeds any single agency's or nation's resources. The commitment to timely, full and open, and unrestricted accessibility to data and information is shaped by the awareness of the complexity of the global environment and an appreciation for the value of observations in understanding and monitoring the Earth system.

ABBREVIATIONS USED

Abbreviations used in the National Strategy and Appendices are listed in alphabetical order.

CENRS	Committee on Environment, Natural Resources, and Sustainability
CEOS	Committee on Earth Observation Satellites
CNES	Centre National d'Edutes Spatiales
DMAC	Data Management and Communications
DOI	Department of the Interior
EOP	Executive Office of the President
ESA	European Space Agency
FGDC	Federal Geographic Data Committee
GCIS	Global Change Information System
GEO	Group on Earth Observations
GEOSS	Global Earth Observation System of Systems
GPS	Global Positioning System
HabCam	Habitat Mapping Camera System
IEOS	Integrated Earth Observation System
ISO	International Organization for Standardization
IOOS	Integrated Ocean Observing System
JWG	NOAA Joint Working Group
NASA	National Aeronautics and Space Administration
NASDA	National Association of State Departments of Agriculture
NEO	National Earth Observations
NOAA	National Oceanic and Atmospheric Administration
NOC	National Ocean Council
NSF	National Science Foundation
NSTC	National Science and Technology Council
OFCM	Office of the Federal Coordinator for Meteorology

OMB	Office of Management and Budget
OSTP	Office of Science and Technology Policy
PCAST	President's Council of Advisors on Science and Technology
QA	Quality Assurance
QC	Quality Control
SBA	Societal Benefit Area
SGCR	Subcommittee Global Change Research
SOA	Service-oriented architecture
SOST	Subcommittee on Ocean Science and Technology
USGCRP	United States Global Change Research Program
USGEO	United States Group on Earth Observations
USGS	United States Geological Survey

End Notes

[1] Non-Federal entities encompass state, regional, local, and tribal governments; nongovernmental organizations; academia; citizen scientists; commercial firms; international organizations; and foreign governments.

[2] See Appendix A.

[3] Non-Federal entities encompass state, regional, local, and tribal governments; nongovernmental organizations; academia; citizen scientists; commercial firms; international organizations; and foreign governments.

[4] Model outputs are generally excluded from this definition; however, some observing systems produce and record measures and observations that may require sensor models to process raw observations to a form in which they are exploitable.

[5] Such an approach is referred to as Integrated Resource Management.

[6] Increasingly, Federal Agencies are relying on observing systems that are obtained in partnership with other entities through co-funding, leveraging, purchase arrangements ("data buys"), and other agreements to provide mutually beneficial Earth observation activities. These systems are often maintained by multiple funding sources and operated by non-Federal entities (domestic and international), either in cooperation with Federal Agencies or independently. Partnership arrangements are particularly common for Earth-observing satellites and other large-scale sensing and observing networks and infrastructure

[7] National security requirements and considerations are not covered by this strategy, though the use of national security assets for civil purposes is included. The Civil Applications Committee is an interagency committee that coordinates and oversees the Federal civil use of classified collections

[8] Group on Earth Observations, Global Earth Observation System of Systems (GEOSS): 10 Year Implementation Plan Reference Document, 2005, www.earthobservations.org/documents/10-Year%20Plan%20Reference%20Document.pdf; Interagency Working Group on Earth Observations, Strategic Plan for the U.S. Integrated Earth Observation System, NSTC Committee on Environment and Natural Resources, 2005, www.whitehouse.gov/sites/default/files/microsites/ostp/eocstrategic_plan.pdf

[9] For the purposes of the assessment, Earth-observing systems were addressed primarily at the program level. Further, consideration of Earth observation systems with primary

dependency on human Earth observation sensing elements were limited to those programs that regularly and systematically collect observations and measures to produce periodically required data that are vital to an agency's mission.

[10] Office of Management and Budget. Circular A-130, Revised (Transmittal Memorandum No. 4), 2000, www.whitehouse.gov/sites/default /files/omb/circulars/a130/a130trans4.pdf.

[11] Committee on Archiving and Accessing Environmental and Geospatial Data at NOAA, Environmental Data Management at NOAA: Archiving, Stewardship and Access, National Research Council, 2007, www.nap.edu/catalog. php?record_id=12017; OSTP Interagency Working Group on Digital Data, Harnessing the Power of Digital Data: Taking the Next Step, 2011; and United States Group on Earth Observations, Exchanging Data for Societal Benefit: An IEOS Web Services Architecture, 2008; U.S. Chief Information Officer, 25 Point Implementation Plan to Reform Federal Information Technology Management, 2010, www.cio.gov/documents/25-Point-Implementation-Plan-to-Reform-Federal%20IT.pdf.

[12] A record schedule is an agency's official policy on how long to retain agency records and how to dispose of them.

[13] See Section 5.4, "Data Sharing" of Group on Earth Observations, "Global Earth Observation System of Systems (GEOSS): 10 Year Implementation Plan," 2005. http://www.earthobservations.org/documents/10-Year%20Implementation%20Plan.pdf.

[14] More information on metadata can be found in Annex D.

[15] The data life cycle discussed in this chapter should not be confused with the concept of life cycle analysis, which is a technique to assess environmental impacts associated with all the stages of a product's development and use from raw material through disposal.

[16] United States Group on Earth Observations, *Exchanging Data for Societal Benefit: An IEOS Web Services Architecture*, 2008, usgeo.gov/images/usgeomain/exchangingdataforsocietalbenefit.pdf

[17] Institute of Electrical and Electronics Engineers, IEEE Standard Computer Dictionary: A Compilation of IEEE Standard Computer Glossaries, New York, NY: 1990.

[18] In this document "data quality" references international standards approaches such as ISO 17025 that include traceability to the International System of Units (SI), the need for measurement uncertainty analysis, and the validation and verification of measurements.

[19] America COMPETES Act (Public Law 110–69), www.gpo.gov/fdsys/pkg/PLAW-110publ 69/content-detail.html.

[20] Congressional direction for this activity is found in language from the National Aeronautics and Space Administration Authorization Act of 2010 (Public Law 111–267): SEC. 702. INTERAGENCY COLLABORATION IMPLEMENTATION APPROACH. The Director of OSTP should establish a mechanism to ensure greater coordination of the research, operations, and activities relating to civilian Earth observation of those Agencies, including NASA, that have active programs that either contribute directly or indirectly to these areas. This mechanism should include the development of a strategic implementation plan that is updated at least every 3 years, and includes a process for external independent advisory input. This plan should include a description of the responsibilities of the various Agency roles in Earth observations, recommended cost-sharing and procurement arrangements between Agencies and other entities, including international arrangements, and a plan for ensuring the provision of sustained, long-term space-based climate observations. The Director should provide a report to Congress within 90 days after the date of enactment of this Act on the implementation plan for this mechanism.

[21] See Annex A.

[22] Group on Earth Observations, *Global Earth Observation System of Systems (GEOSS): 10 Year Implementation Plan Reference Document*, 2005, www.earthobservations.org/documents/10-Year%20Plan%20Reference%20Document.pdf.

[23] Interagency Working Group on Earth Observations, *Strategic Plan for the U.S. Integrated Earth Observation System*, NSTC Committee on Environment and Natural Resources, 2005, www.whitehouse.gov/sites/default/files/microsites/ostp/eocstrategic_plan.pdf.

[24] Office of Management and Budget, *Open Government Directive*, 2009, www.whitehouse.gov/sites/default/files/omb/assets/memoranda_2010/m10-06.pdf.

[25] America COMPETES Reauthorization Act of 2010, (Public Law 111–358), www.gpo.gov/fdsys/pkg/PLAW-111publ358/pdf/PLAW-111publ358. pdf.

[26] National Oceanic and Atmospheric Administration, *NOAA Procedure for Scientific Records Appraisal and Archive Approval: Guide for Data Managers: Guide for Data Managers*, 2008, www.ngdc.noaa.gov/wiki/images /0/0b/NOAA_Procedure_document_final.pdf.

[27] Geo-Data Informatics: Exploring the Life Cycle, Citation and Integration of Geo-Data (Broomfield, Colorado, March 2011), http://tw.rpi.edu/web/Workshop/Community/GeoData2011.

[28] Developing Data Attribution and Citation Practices and Standards (Berkeley, California, August 2011). http:// sites.nationalacademies.org/PGA/brdi/PGA_064019.

[29] The statements from NASA's current Earth Science Data Policy referred to in the body of this plan can be found at science.nasa.gov/earth-science/earth-science-data/data-information-policy/.

[30] CEOS was formed in 1984 by the G-8 governments to foster cooperation among space agencies involved in Earth observing.

[31] *Coordination for the next decade: 1995 CEOS yearbook.* Surrey, England: Smith System Engineering Ltd., 1995.

In: Civil Earth Observations
Editor: Brendon Walters
ISBN: 978-1-63321-965-6

Chapter 2

NATIONAL PLAN FOR CIVIL EARTH OBSERVATIONS*

United States Group on Earth Observations

EXECUTIVE SUMMARY

The U.S. Government is the largest provider of environmental and Earth-system data in the world. These data are derived from observations of the Earth, which are used by Federal agencies and their partners to carry out their missions. These data form the foundation of services that protect human life, property, the economy, and national security, and they support research to foster scientific advances. Provided through public funding, they are made open to the greatest extent possible to advance human knowledge, to enable private industry to provide value-added services, and for general public use.

As the Nation's Earth-observation capacity and related data holdings have grown, so has the complexity of the challenge of managing Earth observation systems effectively and taking full advantage of the data they collect. While Earth observations and data are often collected to support the delivery of well-defined products and public services or meet specific research needs, improved coordination and access would ensure that the data are used more broadly. By expanding the use of observations and data beyond the purposes

---

* This is an edited, reformatted and augmented version of a document issued by the National Science and Technology Council, Office of Science and Technology Policy, July 2014.

for which they are originally collected, the United States can maximize the impact of the resources invested in Earth-observation systems.

In October 2010, Congress charged the Director of the Office of Science and Technology Policy (OSTP) with establishing a mechanism for addressing this challenge through the production and routine update of a strategic plan for Earth observations. In response, OSTP convened a National Earth Observations Task Force (NEOTF) in February 2011, which produced the National Strategy for Civil Earth Observations in April 2013. The NEOTF also conducted the first assessment of the Federal Earth observations enterprise. The resulting Earth-Observation Assessment (EOA) considered the impact of observing systems on distinct societal benefit areas.

This document, the *National Plan for Civil Earth Observations* (hereafter referred to as the National Plan), incorporates the priorities identified in the EOA to provide strategic guidance for a balanced portfolio approach to managing civil Earth observations to fulfill agency mandates and achieve national objectives. As required by law, this National Plan will be updated every three years to provide greater coordination of Federal civil Earth-observation systems.

The National Plan defines a new framework for constructing a balanced portfolio of Earth observations and observing systems. This framework classifies Earth-observation activities according to two broad categories, "sustained" and "experimental" based on the duration of the anticipated Federal commitment:

- Sustained observations are defined as measurements taken routinely that Federal agencies are committed to monitoring on an ongoing basis, generally for seven years or more. These measurements can be for public services or for Earth-system research in the public interest.
- Experimental observations are defined as measurements taken for a limited observing period, generally seven years or less, that Federal agencies are committed to monitoring for research and development purposes. These measurements serve to advance human knowledge, explore technical innovation, and improve services, and in many cases may be first-of-their-kind Earth observations.

Within the subcategory of sustained observations for public services, the National Plan defines two tiers of measurement groups. Tier 1 measurement groups are those derived from systems identified in the EOA as having high impact on a majority of the societal benefit areas; Tier 2 measurement groups include those derived from the remaining high-impact systems. While the EOA provided higher overall scores to Tier 1 systems, many Tier 2 systems contribute critically, or are essential, to key objectives in one or more societal benefit areas. Some Tier 2 systems are the only observing systems available for accomplishing a particular objective.

These new categories advance the Nation's approach to Earth observations by describing a new framework based on the duration of Federal commitment to the period of observation, which is an essential step for prioritizing the Nation's Earth observations portfolio. This framework is also a step toward addressing a key policy challenge in Earth observations: determining when experimental observations should be transitioned to sustained observations for research or for delivery of public services.

Based on this framework and the results of the EOA, the National Plan establishes the following rank- ordered priorities:

1. Continuity of sustained observations for public services
2. Continuity of sustained observations for Earth-system research
3. Continued investment in experimental observations
4. Planned improvements to sustained observation networks and surveys for all observation categories
5. Continuity of, and improvements to, a rigorous assessment and prioritization process

The overall set of observations resulting from these priorities should yield a balanced Earth-observations portfolio.

While the National Plan provides guidance in setting priorities for the construction of the portfolio, agencies have discretion, in consultation with the Executive Office of the President and Congress, to deviate from the National Plan's rankings of priorities when necessary for managing specific systems in the categories and tiers outlined in this document. The National Plan provides this flexibility while still meeting the Nation's overall civil Earth-observation priorities and objectives.

The National Plan also identifies the following rank-ordered supporting actions that will maximize the benefits derived from the Nation's Earth observations:

1. Coordinate and integrate observations
2. Improve data access, management, and interoperability
3. Increase efficiency and cost savings

4. Improve observation density and sampling
5. Maintain and support infrastructure
6. Explore commercial solutions
7. Maintain and strengthen international collaboration
8. Engage in stakeholder-driven innovation

The National Plan also describes specific agency roles and responsibilities for sustaining observation systems and platforms.

Implementation and coordination of the activities outlined in the National Plan will be conducted through the budget and program-planning activities of the relevant Federal agencies and through interagency processes. Federal agencies will determine implementation schedules, progress reviews, and funding profiles in consultation with the Executive Office of the President.

The primary forum for interagency discussion and coordination of Earth observation, related data management, and related international issues is the United States Group on Earth Observations (USGEO) Subcommittee of the National Science and Technology Council (NSTC) Committee on Environment, Natural Resources, and Sustainability (CENRS). OSTP, in consultation with the USGEO Subcommittee, the NSTC CENRS, and their member agencies, will review and update this National Plan on a three-year cycle. As part of the update process, OSTP will solicit and consider the input of external stakeholders and the general public. For this first National Plan, OSTP sought input from external stakeholders through a publicly released Request for Information.

1. Introduction

The U.S. Government the largest provider of environmental and Earth system data in the world. These data are derived from Earth observations[1] collected by numerous Federal agencies and partners to carry out their missions in support of life, property, and economic and national security, and they are the foundation for scientific advances. In accordance with the *National Strategy for Civil Earth Observations* and Executive Order No. 13642, these publicly funded data are made open[2] to the greatest extent possible to advance human knowledge, to enable private industry to provide value-added services, and for general public use.[3]

Conservative estimates indicate that Federal Earth-observation activities add $30 billion to the U.S. economy each year.[4] These investments ensure that decision makers, businesses, first responders, farmers, and an array of other stakeholders have the information they need about natural resources, climate and weather, natural hazards, land-use change, ecosystem health, water, and other characteristics of the planet. Taken together, Earth observations provide the indispensable foundation for meeting the Federal Government's long-term sustainability objectives and advancing U.S. social, environmental, and economic well-being.

As the Nation's Earth-observation capacity has grown, however, so has the complexity and challenge of its most effective use for public benefit. Today, civil Earth observations are funded in the budgets of 11 departments and agencies of the Federal Government, including more than an estimated $2.5 billion in satellite systems and more than $1 billion for airborne, terrestrial (including freshwater), and marine networks and surveys (e.g., buoys, stream gages, and fishery surveys). U.S. Earth observation efforts are distributed among more than 100 programs under the purview of Federal agencies and non-Federal entities that produce and use these data.[5]

While Earth-system data collected through these observations are currently used to meet critical needs of distinct organizations and stakeholders, improved coordination will ensure that information derived from Earth observations will be used more broadly for both traditional and innovative purposes. In October 2010, Congress charged the Director of the Office of Science and Technology Policy (OSTP) with establishing a mechanism for addressing this challenge.[6] OSTP convened a National Earth Observations Task Force (NEOTF) in February 2011 under the National Science and Technology Council (NSTC) Committee on Environment, Natural Resources, and Sustainability (CENRS) to inform the OSTP response to Congress. The NEOTF took three actions:

1. The development of a *National Strategy for Civil Earth Observations* (hereafter referred to as the National Strategy) to provide an enduring framework for routine assessment and planning for the Nation's Earth observation infrastructure. The National Strategy was released in April 2013.[7]
2. The development of a data-management framework, including principles and guidelines to improve discovery, access, and use of Earth observations. This framework is contained within the National Strategy.
3. The first assessment of the Federal Earth-observation enterprise, reviewing the impact of 362 observing systems on 13 societal themes.

Summary results of the 2012 Earth Observation Assessment (EOA) are presented in this National Plan for Civil Earth Observations (hereafter referred to as the National Plan).

These three actions provided OSTP and CENRS with the foundation for this National Plan, which includes the following elements:

1. Definitions and context (Section 2)
2. Categories for civil Earth observations (Section 3)
3. Priorities and supporting actions for civil Earth observations (Section 4)
4. Agency roles and responsibilities for sustained observations from airborne, terrestrial, and marine platforms (Section 5)
5. Agency roles and responsibilities for civil Earth observations from space (Section 6)
6. Summary guidelines on implementation and coordination of the National Plan (Section 7)
7. Summary results from the 2012 EOA supporting the identified priorities, EOA caveats, and a list of abbreviations used in this document (Annexes I-III).

As required by law, this National Plan will be updated every three years to ensure greater coordination of Federal civil Earth observation systems.

This National Plan serves as strategic guidance and sets out to fulfill agency mandates and national objectives via a balanced portfolio approach to civil Earth observations. The National Plan provides a framework that allows for the establishment, evaluation, and evolution of a balanced portfolio of observations and observing systems. This new framework builds on recent progress Federal agencies have made in taking fuller advantage of Earth observations across traditional boundaries to address their mission objectives and policy goals.

2. Definitions and Context

The National Strategy defined key terms and concepts for routine assessment and planning for Earth observations. These and other definitions are used throughout the National Plan.[8]

2.1. Definition of Earth Observations and Earth Observing Systems

"Observation" refers to the act of making and recording the measurement of a phenomenon. "Earth observations" are described in the National Strategy as follows:

> The term "Earth observations" refers to data and products derived from Earth-observing systems and surveys. The term "observing systems" refers to one or more sensing elements that directly or indirectly collect observations of the Earth, measure environmental parameters, or survey biological or other Earth resources (land surface, biosphere, solid Earth, atmosphere, and oceans).[9] Sensing elements may be deployed as individual sensors or in constellations or networks and may include instrumentation or human elements. Observing-system platforms may be mobile or fixed and are space-based, airborne, terrestrial, freshwater, or marine-based.

Earth observations are increasingly provided by integrated systems that support remotely sensed, in situ, and human observations. The benefit of these observations comes from the analysis of Earth-system parameters from different geographic or temporal perspectives, providing more complete monitoring of the target phenomenon and its interaction with other phenomena.

2.2. Societal Benefit Areas (SBAs)

The National Strategy laid out a process to evaluate Earth-observing systems based on the information products and data streams they support in defined SBAs. This approach was adopted by the NSTC CENRS in February 2012, is consistent with the Federal Government's sustainability objectives, and aligns with international agreements and prior interagency work in this area.[10] The first assessment of Earth observations was organized around 13 societal themes, which consisted of 12 SBAs (listed alphabetically below) and the reference measurements that underpin them:

- **Agriculture and Forestry**: Supporting sustainable agriculture and forestry
- **Biodiversity**: Understanding and conserving biodiversity

- **Climate**: Understanding, assessing, predicting, mitigating, and adapting to climate variability and change
- **Disasters**: Reducing loss of life, property, and ecosystem damage from natural and human- induced disasters
- **Ecosystems (Terrestrial and Freshwater)**: Improving the management and protection of terrestrial and freshwater ecosystems
- **Energy and Mineral Resources**: Improving the identification and management of energy and mineral resources
- **Human Health**: Understanding environmental factors affecting human health and well-being
- **Ocean and Coastal Resources and Ecosystems**: Understanding and protecting ocean, coastal, and Great Lakes populations and resources, including fisheries, aquaculture, and marine ecosystems
- **Space Weather**: Understanding, assessing, predicting, and mitigating the effects of space weather on technological systems, including satellites, power grids, communications, and navigation
- **Transportation**: Improving the safety and efficiency of all modes of transportation, including air, highway, railway, and marine
- **Water Resources**: Improving water-resource management through better understanding and monitoring of the water cycle
- **Weather**: Improving weather information, forecasting, and warning
- **Reference Measurements**: Improving reference measurements—the underpinnings of all SBAs— such as geodesy, bathymetry, topography, geolocation, timing, and the fundamental measurement systems and standards supporting them

These SBAs are interconnected at local, regional, national, and international scales and include scientific research, economic activities, and environmental and social domains. Many involve critical government functions, such as the continuity of national government and the protection of life and property.

2.3. Earth Observation Assessment (EOA)

The first EOA was conducted between February and August 2012 under the auspices of the NEOTF. The NEOTF principals designated a working group to collaborate with appropriate subject matter experts (SMEs). OSTP reached out to the subcommittees of CENRS to identify the 13 subject matter leads, who in turn recruited over 300 Federal experts to participate in 26 analytical workshops.

This EOA quantified the impacts of existing observing systems on a set of key objectives defined for each SBA listed in Subsection 2.2. This resulted in the identification of 362 observing systems and surveys, of which 145 were designated as "high impact." Results for the high-impact systems were grouped in tiers and ordered based on a numeric impact score derived through the assessment process. These results are presented in Annex I.[11]

The EOA provided two new perspectives to complement the work of previous studies in this area, namely (a) the inclusion of non-satellite systems and (b) a robust analysis of the impact of each system with respect to its delivery of services to society.[12]

The next EOA, which is planned to begin in 2014, will seek additional insight with regard to research priorities and future needs in addition to existing systems.

2.4. Data-Management Framework, Big-Earth-Data Initiative, and Climate Data Initiative

While Earth observations are typically produced for a specific purpose, they are often useful for purposes not foreseen during their development. Earth observation data can be reused, managed, and preserved such that both anticipated and unanticipated users can find, evaluate, understand, and use the data in new ways to achieve added benefit. The National Strategy, therefore, set out a comprehensive data- management framework to promote improved discoverability, accessibility, and usability of Earth observation data.

The National Plan includes improving data access, management, and interoperability as a supporting action. The Big Earth Data Initiative (BEDI) is designed to support this objective (see Section 4.2.2). In addition, the President's Climate Action Plan[13], announced in June 2013, launched a Climate Data Initiative to leverage extensive Federal climate-relevant data to stimulate innovation and private-sector entrepreneurship in support of climate resilience.

2.5. Relationship between the National Plan and Existing Studies

The National Plan can be understood as the first in a series of interagency efforts to analyze Federal-Earth observation priorities. It can also be understood in connection with other internal and external assessments. Of these assessments, an important example is the 2007 National Academies report, *Earth Science and Applications from Space*, known as the Earth Science decadal survey. This and other reports of the National Academies provide substantial material for understanding Earth-observation priorities of the research community. This National Plan was informed by the results of the EOA, by these reports, and by related interagency deliberation.

2.6. External Input and the Request for Information

In developing this first National Plan, OSTP sought input from external stakeholders through a Request for Information (RFI).[14] Issued in November of 2013, the RFI solicited input on the major themes, categories, and priorities for the National Plan. OSTP received responses from a range of stakeholders, including individuals, academic institutions, private-sector companies, and industry organizations. Using both qualitative and quantitative approaches, OSTP analyzed the RFI responses and incorporated input into the National Plan where appropriate. OSTP will seek and incorporate external input in future editions of the National Plan.

3. Categories for Civil Earth Observations

3.1. Overview of Categories

The National Plan defines two categories of observations that reflect the intention to distinguish systems and programs based on the duration of Federal commitment to the period of observation: either sustained over time or experimental and therefore time-limited. Sustained observations may be used to support public services and research for the public interest. Experimental observations may be used to support a variety of purposes, including: advancing human knowledge through basic and applied research, exploring technical innovation, or improving public services.

A fundamental goal of this National Plan is to achieve a balanced portfolio across and within both sustained and experimental categories of observations. Any Federal agency may engage in sustained and experimental observations to meet its mandate, achieve specific missions, or to support national objectives. These new categories of observation are designed to provide clarity in the Nation's approach to Earth observations. This categorization, based on the duration of commitment, is necessary for prioritizing the Nation's Earth-observing portfolio.[15]

This National Plan recognizes, however, that all civil Earth observations collected by the public sector are considered public goods, and that data from systems in any one category may be reused for purposes other than those for which the observation was originally taken. Such reuse is enabled by the rapid exchange and integration of data made possible by modern information technology (see Subsection 2.4 on data management).

For example:

- Sustained observation systems supporting the delivery of public services contribute significant data and information to both short- and long-term research programs.
- Sustained observation systems for research, as well as experimental observation systems, provide data and information that are routinely exploited in support of the ongoing, regular delivery of public services.
- Experimental observation systems supported by research and development funds can yield new systems or observation capabilities that can then be applied to service-driven observation programs.

While recognizing the value of data across all categories, this approach to categorizing observing systems as either sustained or experimental is a step toward addressing a key policy challenge: determining when experimental Earth observations should be transitioned to sustained observations for research or to help deliver public services. This transition may occur within or between agencies. The policy challenge is greatest in the case of multi-agency collaboration. By recognizing that multiple agencies engage in both experimental and sustained observations, the new terminology for sustained observations allows for long- term measurement responsibilities and plans to be more easily and accurately characterized within individual agency budgets.

This National Plan acknowledges the outstanding need to address cross-agency, experimental, and sustained observation challenges, and it initiates an interagency process for establishing government- wide priorities for sustained observations, either for service or research purposes.

3.2. Sustained Observations

For the purposes of this National Plan, sustained observations are defined as measurements that Federal agencies are committed to taking on an ongoing basis, generally for seven years or more, at a level of quality sufficient for the primary purpose for which the measurement is taken.[16] Such long-term commitments include pre-planned improvements and service-life extension programs. Sustained observations are divided into two purpose-driven subcategories: those for public services and those for Earth system research for the public interest. These subcategories are further described in the following two subsections.

3.2.1. Sustained Observations for Public Services

Sustained Observations for Public Services are those systematic measurements necessary to support products routinely generated for, and widely disseminated to, the general public. These include vital measurements supporting continuous data streams and data products for preservation of life and property (e.g., for severe weather, seasonal and inter-annual climate forecasts, earthquakes, volcanoes, tsunamis, floods, fire detection and suppression, and air quality alerts); routinely generated current conditions data (e.g., for transportation, agriculture, energy, and weather forecasting); and data relating to ongoing resource and environmental management (e.g., for trends analysis, stock assessments, water quality, and forestry).

The Earth-observation systems that produce these sustained observations constitute vital national infrastructure, providing well-established, direct benefits to society and the economy (e.g., protecting life and property and securing food and water during disasters). These systems are in place to provide the Nation with essential information to promote and sustain economic vitality (e.g., transportation, agriculture, energy, water, and natural resource management) and public safety (e.g., hazard warnings). These data streams form the foundation for critical scientific research to improve fundamental understanding of the Earth system and its changing climate. They provide accurate forecasts, surveys, and records that support U.S. scientific, economic, and commercial interests and, as such, are essential to the maintenance of national and economic security.

Sustained observations for public services are subdivided into two tiers, described below.

3.2.1.1. Tier 1

Based on results of the EOA, the following measurement groups support a majority of the societal themes.[17] These measurement groups represent the highest priority measurements in the category of sustained observations for public services. They are listed below in priority order.

1. *Weather and seasonal climate monitoring and prediction:* Observations in this measurement group characterize phenomena such as precipitation, storms, wind, floods, sea state, drought, wildfires, ice, air quality (including ozone), and weather risks to human health and transportation. They also contribute to short-term climate forecasts. These observations derive from next- generation radar on fixed and mobile platforms; atmospheric sounding from space and airborne platforms, oceanic measurements, and spectral and radiometric imaging of the Earth surface.
2. *Dynamic land-surface monitoring and characterization:* Observations in this measurement group support food and water security, water availability and quality, fire detection and suppression, human health, forestry, soil characterization (including soil moisture), hazards mapping and response, and natural-resource management. They particularly derive from multispectral and hyperspectral imaging from space and airborne platforms, forest inventory, and soil data collection.
3. *Elevation and geo-location:* Observations in this measurement group support food and water security, hazard and risk mapping, and natural-resource management. These observations particularly include topography and bathymetry, surface modeling, hydrologic data, and ecosystems-related data as derived from radar and laser sensors on satellite-based, airborne, and terrestrial platforms, as well as positioning, navigation, and timing satellites, such as those used for the Global Positioning System (GPS).
4. *Water level and flow:* Observations in this measurement group support coastal inundation and inland flooding, water availability, hydropower management, transportation, human health, water equivalent of snow, and tsunami hazard preparedness. They particularly derive from coastal and inland water level and flow

measurements, seafloor topography, and ocean topography from satellite altimetry.

3.2.1.2. Tier 2

The following measurement groups are identified based on the remaining high-impact observing systems identified in the EOA.[18] These measurement classes are of next-highest priority and importance in the category of sustained observations for public services.[19] They are listed here in alphabetical order.

- *Ecosystem and biodiversity resource surveys* for terrestrial, freshwater, and marine ecosystems, including fisheries and wildlife management
- *Environmental-quality monitoring*, specifically disease-vector surveillance, water quality, and air quality associated with changes in atmospheric composition, including particulate matter and short-lived climate pollutants
- *Geo-hazard monitoring* for earthquakes, volcanoes, landslides, regional and local subsidence (e.g., sinkholes), inundation, and tsunamis
- *Space-weather monitoring* of geomagnetic storms, sunspots, solar flares, associated x-ray and ultraviolet emissions, solar wind (including coronal mass ejection), solar energetic particles, traveling ionosphere disturbances, and associated changes of the Earth's geomagnetic field and ionosphere for their impact on human activities

While the EOA provided higher overall scores to those systems that were identified as impacting more than one SBA, some of these systems contribute critically or are essential to key objectives in one or more SBAs. Some Tier 2 systems are critical in that they are the only observing systems available for a particular objective, so that objective could not be achieved without them.

3.2.2. Sustained Observations for Earth-System Research in the Public Interest

The public interest also requires sustained observations for understanding how and why the Earth system, including the Earth's climate, is changing. These observations are those measurements supporting continuous data streams or routinely generated data products that are needed for basic and applied research to advance human knowledge (climate-change research, solid-Earth research, meteorological research, ocean and water-cycle research, and space-weather research), to improve public services, and to support public and general education. These observations often require multi-year data collection and maintenance within a specific sampling frame (e.g., measurements taken at a specific location at a given interval). The purpose of such sustained observations is often long-term research, but the data collected often have immediate benefit for society and are frequently integrated into sustained services.

The following measurement categories, presented in alphabetical order, were identified for sustained research observations.[20]

- *Atmospheric state*, including measurements of temperature, pressure, humidity, wind, and ozone at the accuracy required for long-term climate research, and, as appropriate, to improve short- and medium-range weather forecasting;
- *Cryosphere*, including measurements of ice sheets, glaciers, permafrost, snow, and sea ice extent and thickness;
- *Earth's energy budget*, including total solar irradiance and Earth's radiation budget, and the reflectance and scattering properties of clouds, aerosols, and greenhouse gases, specifically for understanding Earth's sensitivity to climate change;
- *Extremes*, including specific and routine observations for the study of extreme temperatures, drought, precipitation, and wind;
- *Geo-hazard research*, including monitoring land-surface deformation to better understand regional and local disaster potential and effects, and the monitoring of phenomena that precede natural disasters, such as seismic, stress, strain, and geochemical and temperature changes;
- *Greenhouse gas emissions and concentrations*, including understanding sources and sinks of greenhouse gases, as well as changes in long-lived greenhouse gas and short-lived climate-pollutant concentrations over time;
- *Integrated geophysical and biosphere characterization (terrestrial, freshwater, and marine)*, including long-term dynamics to understand ecosystem change and biogeochemical processes, particularly the carbon cycle;
- *Ocean state*, including observations of sea levels, temperature, salinity, pH, alkalinity, currents and characteristics of marine ecosystems;
- *Space weather*, including long-term understanding of the Earth-Sun relationship, solar dynamics, and the drivers of space-weather impacts

at the Earth's surface, such as coupling between space weather and geomagnetic storms; and,
- *Water cycle*, including the analysis of droughts, floods, and water availability (precipitation, soil moisture, snow-water equivalent, evapotranspiration, groundwater, surface water, and runoff).

3.3. Experimental Observations

Experimental observations are defined as measurements planned for limited durations, generally seven years or less, that Federal agencies are committed to making for limited research and development purposes. These observations may be taken for a variety of purposes: to advance human knowledge through basic and applied research, to explore technical innovation, or to improve public services. These include first-of-their-kind observations, technology innovation and infusion (e.g., new methods, proofs of concept, and evolving observation platforms such as small satellites, microbuses, unmanned aerial vehicles, automated mobile and distributed sensor networks, and handheld devices), targeted activities (e.g., field campaigns, process studies, model validations, and design studies), and experiments for system improvements (e.g., risk reduction and upgrades).

The advancement of experimental scientific research on the Earth's fundamental processes is central to human progress, applicable to improvements in current and future service capabilities, and essential for achievement of other national Earth science objectives. Therefore, the Federal Government will pursue experimental and first-of-their-kind Earth observations to advance human knowledge, explore technical innovation, and improve services.

Innovation is an essential component in developing a robust portfolio of experimental observation systems. Evolving observation platforms offer the potential to make Earth observations more efficient, accurate, or economical over the National Plan's three-year timeframe. The infusion of new technologies requires investments in new sensors, materials, techniques, satellite architectures, and so forth, supported by research and development funding and combined with new scientific research with the objective of developing more efficient or higher quality observations. Experimental measurements drive innovation for research, form the basis of planned improvements in service-driven observations, and advance human knowledge and understanding.

Within this measurement category, the large-scale Earth science research programs of the National Aeronautics and Space Administration (NASA), the National Science Foundation (NSF), the National Oceanic and Atmospheric Administration (NOAA), the United States Geological Survey (USGS), the Department of Energy (DOE), the United States Department of Agriculture (USDA), the Department of Transportation (DOT), and other agencies will continue setting priorities and advancing innovation in Earth observation by selecting and funding experimental observations.[21]

4. PRIORITIES AND SUPPORTING ACTIONS FOR CIVIL EARTH OBSERVATIONS

Federal agencies will work with the Executive Office of the President and Congress to consider the following priorities and supporting actions. All priorities and activities outlined are to be conducted when feasible and within approved agency budgets.

4.1. Priorities

The most important principle governing the Nation's civil Earth observing systems is that the overall set of observations must yield a balanced portfolio across both sustained and experimental observation. This National Plan provides guidance in setting priorities for the construction of this balanced portfolio. In the balanced framework, no one category, tier, or observation automatically supersedes or supplants all others. Agencies, in consultation with the Executive Office of the President and Congress on these priorities, have discretion to deviate from this National Plan's rankings when necessary. In recognition of this, the National Plan accommodates this flexibility while meeting the Nation's overall civil Earth observation priorities and objectives.

The framework allows for, and recognizes the importance of, planned improvements to observing systems to maintain data of sufficient quality, while the Nation also seeks to improve observation techniques and the effectiveness and efficiency of Earth observations collection, dissemination, and use.

The priorities are listed in ranked order and are elaborated on in the subsections that follow.

1. *Continuity of sustained observations for public services.* Prioritize system investments to ensure continuity of public services.
2. *Continuity of sustained observations for Earth system research.* Establish and maintain programs to ensure data continuity for high-impact sustained research observations.
3. *Continued investment in experimental observations.* Continue to invest in research and development, incorporating technological advances to improve observations.
4. *Planned improvements to sustained observation networks and surveys for all observation categories.* Proceed with planned improvements to sustained observation systems.
5. *Continuity of, and improvements to, a rigorous assessment and prioritization process.* Initiate a national-level process for prioritizing sustained observations for both research and public services and for experimental observations, including a process for external advisory input and strategic balance.

4.1.1. Priority 1: Continuity of Sustained Observations for Public Services

Federal agencies engaged in Earth observations should work both individually and collectively to ensure the continuity of sustained observation systems essential in the maintenance of public safety, national and economic security interests, and programs supporting services critical to agency missions and scientific research. In particular, the President's FY 2015 budget provides support for Federal agencies to:

- Maintain the continuity of observations of current atmospheric conditions from satellites, terrestrial networks, and airborne and marine platforms to provide sufficient weather, hazard, and air-quality forecasting and prediction; the efficient movement of commerce and goods; and the reanalysis of data sets for climate research and infrastructure planning.
- Maintain a sustained, space-based, land-imaging program while ensuring the continuity of 42 years of multispectral information and 36 years of thermal-infrared land-surface information from space, which are unique sources of terrestrial data for understanding land cover change.
- Maintain sufficient continuity of airborne remote-sensing capabilities for agricultural resource monitoring, as well as sustained observations

from multiple platforms for forest inventories and soil surveys. Together these observations are a critical source of information on the management of Earth resources needed to sustain human life, agriculture, forestry, and economic activity.

- Ensure the continuity of the GPS satellite network and maintain related terrestrial reference- frame measurements. GPS is singularly important as the principal and irreplaceable reference for universal time and geo-reference measurements that underpin nearly all Earth observations.
- Establish and maintain a national program to standardize the regular collection of nationwide, high-resolution, three-dimensional data for surface modeling and volumetric analysis for multiple requirements (e.g., airborne light detection and ranging or LIDAR).
- Ensure continuity of terrestrial, marine, coastal, and inland water-level and flow measurements and maintain the continuity of ocean and seafloor topography measurements from satellite altimetry.
- Maintain continuity of space-weather observations, including sustained space-weather research measurements and ground observations, to provide definitive space weather forecasts, warnings, and alerts to the general public, industry, and government agencies.
- Maintain sufficient continuity of data from oceanic platforms (e.g., gliders and moored and drifting buoys).
- Maintain sufficient continuity of observations for ecosystem-resource surveys, environmental- quality measurements, and monitoring of solid-Earth hazards.

Current observing systems supporting this priority are shown in Annex I. To maintain continuity of these observations, follow-on and experimental observation systems will be developed within the timeframe of this National Plan.

4.1.2. Priority 2: Continuity of Sustained Observations for Earth System Research

Federal agencies will continue to collaborate with each other and with international partners to establish and maintain systems and networks specifically for sustained observations for Earth-system research.

Multiple agencies maintain extensive Earth-science programs that will continue to identify priorities for sustained observations for use in Earth-system research. The agency processes that direct these research programs will include consultation with external stakeholders, such as the National Academies, as appropriate. The President's FY 2015 budget provides specific support for agencies and their research programs to continue and strengthen their activities in support of this priority.

For example:

- NASA is extending measurements of groundwater, greenhouse gases, and aerosols from space- based platforms, in collaboration with international partners (see Section 6).
- NOAA is supporting a nationwide Federal network specifically designed and operated for the purpose of long-term climate monitoring. It primarily takes highly accurate measurements of air temperature, precipitation, soil temperature, moisture, and relative humidity for the purpose of detecting long-term climate change on a national scale. It also takes secondary or ancillary measurements of solar radiation, infrared temperature, and wind to assist in the calibration of the primary variables.
- NSF is supporting a federally funded, multi-site, national network designed to gather and synthesize data on the impacts of climate change, land-use change, and invasive species on natural resources and biodiversity. The network will combine site-based data with remotely sensed data and existing continental-scale data sets (e.g., satellite data) to provide a range of scaled data products that can be used to describe changes in the Nation's ecosystem over time.
- NSF has begun construction on a federally funded sensor network to measure the physical, chemical, geological, and biological variables in the ocean and on the seafloor. Knowledge of these variables improves detection and forecasting of environmental changes and their effects on biodiversity, coastal ecosystems, and climate.

Finally, multiple agencies are supporting the development of multi-site and multi-platform observing systems and networks that integrate observations on local to global scales to improve climate-change, biodiversity, and ecosystems research, among other areas.

4.1.3. Priority 3: Continued Investment in Experimental Observations

Experimental observations strengthen the Nation's Earth observation activities through the development and validation of new science, technologies, systems, techniques, and measurements to support both services and scientific research. The continual integration of experimental observations into the portfolio of Federal Earth-observing systems ensures that cutting-edge capabilities are maintained and measurements of increasing quality, accuracy, resolution, and density are provided. The President's FY 2015 budget provides funding for Federal agencies to act individually and in collaboration to support this priority.

4.1.4. Priority 4: Planned Improvements to Sustained Observation Networks and Surveys for All Observation Categories

Federal agencies will proceed with planned improvements to sustain observation systems, with special attention and priority given to include agencies' pre-planned improvements and service-life extension programs.

4.1.5. Priority 5: Continuity of, and Improvements to, a Rigorous Assessment and Prioritization Process

Federal agencies will collaborate with external stakeholders to continue and improve the assessment process initiated by the National Strategy. To this end, the newly re-chartered United States Group on Earth Observations (USGEO) Subcommittee of the NSTC's CENRS will oversee the EOA process on three-year cycles, with the next EOA beginning in 2014. The next EOA will also seek additional insight with regard to research priorities and future needs.

4.2. Supporting Actions

The supporting actions required to meet the foregoing priorities are listed below in priority order and elaborated on in the subsections that follow.

1. *Coordinate and integrate observations.* Coordinate observing and integrate separate observations from multiple platforms, as appropriate, to include federated data sharing standards, ontologies, and user-adopted conventions.
2. *Improve data access, management, and interoperability.* Improve data discovery, access, and use of Earth observations by making them interoperable, providing open access to them, and presenting them in a machine-readable form for end user applications and services.
3. *Increase efficiency and cost savings.* Increase efficiency and cost savings of observation systems, by exploring tradeoffs among cost,

capabilities, and risk and, as appropriate, reevaluating program overhead and management structures.

4. *Improve observation density and sampling.* Where appropriate and cost-effective, upgrade observing systems and widen their data dissemination to improve Earth-observation capabilities and reduce gaps in coverage.
5. *Maintain and support infrastructure.* Maintain necessary infrastructure to operate, manage, house, transport, deploy, modify, and support needed observing systems.
6. *Explore commercial solutions.* Improve entry points for exploiting cost-effective commercial solutions for the provision of Earth observations to encourage private-sector innovation and services while preserving the public-good nature of Earth observations.
7. *Maintain and strengthen international collaboration.* Maintain and strengthen international collaboration and data sharing.
8. *Engage in stakeholder-driven data innovation.* Engage with the private sector and the general public to encourage innovations for collection, exploitation, and wider use of Earth observations based on improved availability of open data, including new applications, new services, citizen science, and crowdsourcing.

4.2.1. Action 1: Coordinate and Integrate Observations

Sustained and experimental observations serve both services and research needs. Many near-term service benefits flow from research observations. Similarly, research is advanced through the data collected by service-driven observations that are archived, disseminated, and reused beyond the immediate user community. Agencies will examine improved methods for integrating sustained and experimental observations, and will explore collaboration to:

- Develop integrated Earth-observation requirements, recognizing unique requirements as appropriate.
- Advance algorithm development for integrating Earth observations from multiple sources, including improved capacity to integrate data from multiple spatial scales and reference models.
- Develop specific, coordinated observing systems and programs to meet integrated requirements; in particular, explore increased airborne, terrestrial, and marine data-collection coverage, coordinated with planned new satellite-observation systems.

- Enhance systematic strategies that can be used for calibration and quality assurance of all observation systems, whether airborne, terrestrial, or marine and satellite based, and develop standardized methods for collection of automated sensor-derived and human-acquired measurements.
- Consolidate, as appropriate, existing networks and mapping through interagency agreements that ensure interoperability, common measurement standards, and meaningful leveraging of resources. Examples of this approach include:
  - Bringing current coastal in situ systems together into a single, federated, coastal observing system under an optimized plan for the U.S. Integrated Ocean Observing System (IOOS®) and
  - Consolidating various land, coastal, and ocean ecosystem mapping projects into a national ecosystem map.
- Consolidate GPS surface-monitoring stations by reducing the number of stations where feasible without reducing coverage. Examples of GPS station networks include those used:
  - Through NSF for crustal motion detection,
  - By USGS for earthquake and volcano-hazard monitoring and early warning, and
  - By NOAA for tropospheric sounding activities.
- Continue development and operation of multi-platform observing systems that combine space, air, land, or ocean measurements to provide cost-effective, integrated analysis of Earth science phenomena.
- Coordinate efficient spectrum allocation and use for observing systems to enable the transmission of large amounts of data, particularly from Earth observing satellites.

4.2.2. Action 2: Improve Data Access, Management, and Interoperability

Realizing the full benefit of the nation's Earth observation investments requires more than the effective deployment of sensors and surveys. Observation data streams and their metadata generally feed machine-to-machine data processors that analyze, automate, integrate, display, and provide decision support tools to make use of large volumes of base data. By improving the management and preservation of Earth-observation data, Federal agencies help to ensure that both anticipated and unanticipated users can find, obtain, evaluate, understand, compare, and use legacy data in new ways.

Under this supporting action, Federal agencies will strengthen their individual efforts and collaborations to improve data discovery, access, archives, and use of Earth observations by making them interoperable, open, subject to shared generic message exchange patterns (e.g., publication/subscription), and machine-readable.[22] To this end, Federal agencies will support development of a comprehensive data-management framework as an integral element of Earth-observation activities to improve access to, and use of, Earth observations by new and potential users. To accelerate the implementation of this framework, the President's FY 2014 and FY 2015 budgets invest in the Big Earth Data Initiative (BEDI) for standardizing and optimizing the collection, management, and delivery of Federal civil Earth-observation data.[23] BEDI also builds upon data-management work already occurring at the agencies.

Federal agencies will collaborate to do the following:

- Promote implementation of end-to-end life-cycle data management, including data-management principles, guidance, and data policy; life-cycle data-management approaches; architectural considerations; standardization; data evaluation; and data-telecommunication efficiency and innovation.
- Encourage the development and use of uniform methodologies and practices across Federal agencies for common services in the handling of Earth-observation data to increase interoperability through improved metadata standardization, filter, and subset services based on user needs, federated user management efforts, and robust standards for multi-use observation systems like LIDAR that are used across various Federal and State agencies.
- Maximize the likelihood that Earth observations are made known, made available, and disseminated to users in a timely and useable manner.
- Continue support for data-clearinghouse mechanisms that emphasize metadata cataloging (such as Data.gov) and advocate support for developing "big data" initiatives and common support services that aim to improve the discoverability, accessibility, usability for intended purpose, and re-purposing of the vast amounts of data already available.
- Support the further development of forecast models, sensor integration, display applications, signal-processing algorithms, and other data-collection and management techniques.

- Work within current, or support the development of new, communities of practice focused on data integration and cross-program coordination to address:
    - Community-based data standards;
    - Web services;
    - Application development;
    - Design of interoperable data systems based on technical standards;
    - Reduction of uncoordinated and duplicative web portals;
    - Development of standardized data products; and
    - Use of cloud-based infrastructure, platforms, and services for efficient data storage and virtual resource sharing.
- Create frameworks, platforms, and systems to integrate data obtained from multiple observations that measure common phenomena.
- Consider improving interoperability between Earth observation data and relevant data sets that are not based on Earth observations (e.g. census data) to support agency objectives.

4.2.3. Action 3: Increase Efficiency and Cost Savings

Federal agencies shall increase efficiency and cost savings in observation systems by exploring tradeoffs among cost, capabilities, and risk and, as appropriate, evaluating program-management structures and overhead. Specifically, agencies will take steps as follows:

- Increase efficiency.
    - Develop specific processes for regular evaluation of in situ airborne, terrestrial, marine, and satellite-based observation systems, both mobile and fixed. These evaluations should be used to guide the consolidation, relocation, expansion, or reduction of sites, instruments, or measurements when such actions offer improved efficiencies or cost-savings while maintaining data quality.
    - Evaluate commercial and foreign satellites and observation networks for their potential value in augmenting the U.S. Government's Earth-observing capacity.
    - Explore possibilities for common ground systems and dissemination mechanisms for satellite- derived information.

- Consider refining satellite risk profiles for nonhuman spaceflight and streamlining review cycles to achieve cost savings and efficiencies, as appropriate.
- Explore coordinated acquisition of sonar, radar, and LIDAR technology, as appropriate, to achieve efficiencies.
- Conduct a study of radiosonde observations in proximity of aircraft soundings to achieve savings for the purpose of weather forecasting.
- Increase coordination and cooperation in the use of vessels for oceanographic research and surveys of living marine resource.
- Explore opportunities to leverage land- and ocean-observing platforms across programs and agencies.
- Use modeling tools to supplement monitoring networks to fill in spatial and temporal gaps and potentially reduce monitoring requirements.
- Explore greater use of National Technical Means resources that could fulfill societal and research requirements through civilian tasking, unclassified product development, and public distribution.

- Improve coordination of sensor development and acquisition of imagery and data.
  - Improve interagency and external coordination in the organization of space-based sensor development and launch.
  - Develop a collaborative strategy among agencies and their primary users on climatology and continuity for ocean-color calibration, data processing, and product development through satellite platforms.
  - Coordinate airborne research efforts among interagency and international partners.
- Increase agencies' automation, remote configuration, and standby mechanisms.
  - Explore automation of surface-air-quality networks and connection with networks for other purposes to increase data utility across observation systems and programs.
  - Explore and, where cost-effective and appropriate, expand the use of airborne, terrestrial, or marine autonomously or remotely operated vehicles with sensor packages to provide more agility and to reduce costs in data acquisition.

- Adopt automated measurement and data-collection techniques as legacy networks are modernized.
- Adopt common software environments to allow for remote configuration updates of surface- observing platforms that will increase efficiency and reduce the cost of retrofitting individual sites.
- Increase coordination of systems making compatible observations to enhance continuity of services and increase cost-savings. This might include optimizing the frequency and density of observations from systems in proximity to maximize coverage of specific phenomena (e.g., severe storms).

- Improve technical refresh for cost reduction.
  - Pursue improvements through value engineering and technical refresh activity to reduce life- cycle maintenance costs, eliminate proprietary restrictions, maintain replenishment sources when the original equipment manufacturer ceases support, and prevent information- technology obsolescence.

4.2.4. Action 4: Improve Observation Density and Sampling

Federal agencies will explore and pursue upgrading systems and techniques through improvements to spatial resolution, temporal cycle, sample density, and geographic coverage of observation networks to close coverage and performance gaps and improve calibration and validation of measurements. When feasible and within approved agency budgets, Federal agencies will, individually or collaboratively:

- Sustain and improve spatial, spectral, and temporal resolution.
  - Collect and employ Earth-observation data, imagery, and reference measurements with higher temporal, spatial, and spectral resolution through more sustained observations and better reuse of related data, particularly for land imaging, boundary layer observations, air- quality measurements, natural hazards, hydrology, and severe-weather forecasting.
- Improve observation density and sampling through technical upgrades.
  - Upgrade receivers to take advantage of improved positioning, navigating, timing, and geodetic-observation capabilities as they are launched to provide better results and increased coverage.

- Pursue the systematic integration and improvement of observing-system density and sampling with hydrodynamic and inundation models through use of uncertainty analyses to identify areas requiring new observations and better model performance.
- Invest in advanced data calibration to improve real-time quality control, including field equipment, test kits, self-calibrating sensors, and other novel technologies.
- Explore technologies that facilitate new measurements or greatly enhance current measurements, such as biotechnology for genetic user identification, portable laboratories and chips, and observer-centric mechanisms.
- Equip and modify existing or planned platforms for sustained, multipurpose observations as appropriate.

- Create new data sets from forthcoming systems (e.g., air quality from geostationary environmental satellites and flight trajectory-oriented weather data from air transportation systems).

4.2.5. Action 5: Maintain and Support Infrastructure

Sustained Earth observations require the operation and maintenance of extensive physical, cyber, communications, and human infrastructure. Such infrastructure supports the maintenance, deployment, retrieval, replacement, and repair of Earth observation systems.[24] Agencies will identify, prioritize, and implement activities and document appropriate funding requirements for life-cycle operation, maintenance, and evolution of infrastructure required for them to sustain Earth-observation systems. As part of this process, agencies shall also consider cost-sharing for complementary and common infrastructure in support of their Earth-observation objectives.

Furthermore, agencies often conduct observations in remote areas and at great distances. The necessary supply-chain infrastructure to support these observations includes the Federal oceanographic and airborne research fleets and other systems. These fleets support sustained observations for both services and research across a broad spectrum of national needs.[25] Agencies should continue to maintain their relevant supply-chain infrastructure and document their related funding requirements.

4.2.6. Action 6: Explore Commercial Solutions

Federal agencies will identify and pursue cost-effective commercial solutions to encourage private-sector innovation while preserving the public-good nature of Earth observations. U.S. agencies will consider a variety of options for ownership, management, and utilization of Earth observation systems and data, including managed services (Government-Owned/ Government-Operated, Government-Owned/Contractor-Operated, or Contractor-Owned/Contractor-Operated), commercially hosted payloads, commercial launch, commercial data buys, and commercial data management. In developing such options, agencies will preserve the principles of full and open data sharing, competitive sourcing, and best value in return for public investments within legal and financial constraints.

4.2.7. Action 7: Maintain and Strengthen International Collaboration

The global nature of many Earth observations and the value of these observations to U.S. Government decision makers require U.S. agencies to carry out their missions through collaboration with foreign agencies, international organizations, and standards/coordination groups. Through international collaboration, U.S. agencies leverage foreign data and scientific expertise to improve their understanding of remote areas, such as the open ocean and polar regions, and to characterize global atmospheric, oceanic, and terrestrial phenomena. In addition, collaboration with international partners helps to minimize unnecessary redundancy in the collection of Earth observations, and ensures the effective use of limited resources. U.S. agencies also work closely with the Department of State and other agencies to provide associated scientific and technical support for U.S. foreign policy, security, economic, and environmental interests.

The concept of integrated Earth observations and open data management achieved international prominence in 2005 with the establishment of the intergovernmental Group on Earth Observations (GEO) and its agreement to support the development of a Global Earth Observation System of Systems (GEOSS).[26] This concept has been endorsed and is being implemented by over 90 governments, the European Commission, and more than 60 international organizations. The United States co-leads this international activity with China, the European Commission, and South Africa. GEOSS offers a comprehensive approach to observe all aspects of the Earth system and integrate the data gathered from these observations into timely and useful information for all sectors of society.

U.S. agencies engage in bilateral and multilateral collaboration with foreign partners and international organizations to obtain Earth-observation measurements necessary to improve U.S. scientific research, environmental monitoring, and related public- and private-sector decision making. Through their international activities, U.S. agencies support full and open data exchange and collaborative research on matters of national and global importance.

In support of this action, Federal agencies will individually or collaboratively take the following actions:

- Maintain and expand bilateral and multilateral relationships as appropriate in support of national and international objectives.
- Improve access to non-U.S. satellite data, especially for domestic users.
- Explore the development of mechanisms for optimizing the collection, processing, and archiving of satellite data with the European Space Agency, the Japan Aerospace Exploration Agency, and other agencies to address specific SBAs (e.g., water).
- Continue U.S. involvement in global observation efforts and national activities that serve to further integrate national observing programs with internationally networked systems.
- Assess the potential for joint collection and processing of data from land-imaging systems to achieve savings through consolidation of image processing and to create a global source of timely, routinely available, ready-to-use measurements.
- Increase the level of international satellite collaboration to obtain specific measurements, following examples set by current collaboration in areas such as sea level, soil moisture, greenhouse gases, and ozone.

4.2.8. Action 8: Engage in Stakeholder-Driven Data Innovation

To improve the collection, exploitation, and use of Earth observations, Federal agencies will pursue private-sector innovations, public data crowdsourcing, and citizen science.[27] These three observation sources are all key components of a portfolio approach to Earth observations. Expected future technical developments throughout the National Plan's lifetime mean crowdsourcing, citizen science, and collection from individual mobile platforms will become increasingly important as they mature. Federal agencies will explore, support, and pursue efforts to enable stakeholder-driven data innovation from Earth observations.

Several initiatives have been launched to foster innovation in the exploitation and use of Earth observation data in the Climate SBA. Consistent with Executive Order No. 13642, "Making Open and Machine Readable the New Default for Government Information," and part of *The President's*

Climate Action Plan, the Obama Administration launched a Climate Data Initiative in March 2014. This initiative leverages extensive Federal climate-relevant data to stimulate innovation and private-sector entrepreneurship in support of national climate-change preparedness. In conjunction with this initiative, Federal agencies will create a virtual toolkit that provides access to data-driven climate resilience tools, services, and best practices, including those developed through the Climate Data Initiative. Federal agencies should continue support for these and other initiatives encouraging innovative use of Earth observation data.

5. Civil Earth Observations from Airborne, Terrestrial, and Marine Platforms

Multiple Federal agencies manage the U.S. Government's airborne, terrestrial (including freshwater), and marine civil Earth-observation platforms and surveys, which gather remotely sensed, in situ, and human observations. The agencies that conduct or fund these observations include the Departments of Agriculture, Commerce, Defense, Energy, Interior, and Transportation; the Environmental Protection Agency (EPA); NASA; NSF; the Smithsonian Institution; and the U.S. Agency for International Development. In general these agencies, in accordance with their existing legal authorities, will continue to pursue the development of these platforms to observe and study the Earth system, with a focus on measurements described in this National Plan.

Airborne, terrestrial, and marine observations are vital to fulfilling Federal public-service obligations and research objectives across multiple SBAs. They provide critical information at high degrees of resolution, density, and efficiency. In addition to serving as primary sources of critical measurements and observations for public services and research across SBAs, these observations are essential to validate satellite-derived data products. Sustained airborne, terrestrial, and marine observations support the provision of public services and scientific research, while experimental observations in each of these categories both advance the state of science and improve the ability to monitor and measure the Earth system.

Continuous high-quality observations are critical for defining the current state of the Earth system; in particular, the constantly changing conditions of the atmosphere, hydrosphere, and biosphere. Observations from airborne, terrestrial, and marine platforms are required to accurately measure a number of Earth-system processes, including those related to biodiversity, groundwater, carbon sequestration, and the subsurface ocean. Long-time-series data derived from these observations contribute to more effective detection and diagnosis of climate change. Agencies should sustain the operations of established airborne, terrestrial, and marine observation platforms with ongoing attention to sufficient coverage and data quality.[28]

Recommendations from teams of SMEs for each SBA, as contained in the 2012 EOA, emphasized the need for sustained observations across SBAs to mitigate the significant risk of observation and data gaps, and, where possible and within agency approved budgets, the need for maintaining and improving the coverage and density of airborne, terrestrial, and marine observations, systems, and programs to preserve and enhance the long-term record.

The following categories highlight examples of high-impact airborne, terrestrial, and marine platforms and programs identified in the EOA. They provide sustained observations for public services, research, and experimental observations. Many of these observing systems are operated in partnership between Federal and non-Federal entities and many Federal agencies rely on these data for their operations. These observations are grouped by platform and are presented in alphabetical order within each category (sustained and experimental).

5.1. Sustained Airborne, Terrestrial, and Marine Observations for Public Services and Research

5.1.1. Airborne Observations for Public Services and Research

The following examples of high-impact airborne platforms and programs identified in the EOA are presented alphabetically.

- *Airborne light detection and ranging (LIDAR):* Airborne observations used to create very high- resolution maps and elevation data sets, among other applications, are a high-impact category of observations conducted by multiple Federal and State agencies, commercial providers, and international partners. The 2012 EOA contained multiple recommendations from SME teams for sustained, expanded airborne three-dimensional measurements, which would benefit multiple SBAs. These measurements would significantly enhance high-resolution digital elevation models and shoreline mapping to improve services in the Disasters and Ocean and Coastal Resources and Ecosystems SBAs.

- *Airborne meteorological data collection and reporting:* Aircraft-based meteorological observations support improved weather forecasting, particularly for upper-air winds and severe weather. Real-time automated position and weather reports are used in predictive weather models on a daily basis. These measurements have an impact on eight SBAs, with "highest" impact on the Weather SBA.
- *Digital orthophotography*: Aircraft-based digital orthophotography provides imagery of the continental United States during agricultural growing and non-growing seasons. These measurements have an impact on nine SBAs, with "very high" or "highest" impact in the Agriculture and Forestry, Biodiversity, and Ecosystems SBAs. Agriculture SMEs recommended increasing the consistency of this imagery collection and integrating it with other relevant observations.
- *Radiosonde observations:* Radiosondes collect information on atmospheric temperature, pressure, and humidity using an instrument suspended from a balloon. The Federal Government has conducted upper air observations with radiosondes since the 1930s and currently uses radiosonde data in weather-prediction models; local severe storm, aviation, and marine forecasts; air-pollution models; satellite-data verification; and analysis of climate variability and change. Radiosonde observations impact nine SBAs, with "very high" impact in the Transportation SBA and "high" in Weather SBA.

5.1.2. Terrestrial (Including Freshwater) Observations for Public Services and Research

The following examples of high-impact terrestrial platforms and programs identified in the EOA are presented alphabetically.

- *Ground-based Weather Radars*: A national ground-based network of weather radars supports weather forecasting and warning services. These systems detect precipitation and wind and contribute to severe-weather and flash-flood warnings, air-traffic safety, flow control for air traffic, resource protection at military bases, and management of water, agriculture, forests, and snow removal. The network impacts nine SBAs, with "highest" impact in the Transportation SBA and "very high" impact in the Weather and Energy and Mineral Resources SBAs.

- *Soil Observations:* A nationwide partnership of Federal, regional, State, and local agencies, along with private entities and institutions, cooperatively investigates, inventories, documents, classifies, interprets, disseminates, and publishes information about soils of the United States and its territories. These activities are carried out at national, regional, and State levels. This information has a "high" impact in the Agriculture and Forestry, Biodiversity, Ecosystems, Human Health, and Water Resources SBAs.
- *Stream Gage Network:* A nationwide network of stream gages operated by Federal, State, and local agencies provides stream flow monitoring and measurement and hydrological observations. These observations provide critical support to stream-flow forecasts, river-basin outflow forecasts, drought forecasts, water-quality measurements, and sentinel watershed monitoring. The network impacts ten SBAs, with "very high" impact in the Water Resources and Transportation SBAs.

5.1.3. Marine Observations for Public Services and Research

The following examples of high-impact marine platforms and programs identified in the EOA are presented alphabetically.

- *High-Frequency Coastal Radar Network:* An integrated network of high-frequency coastal radars, that is operated by Federal and institutional partners and provides measurements of surface currents over wide geographical areas, supporting the Transportation SBA.
- *Oceanic Buoys and Coastal Networks:* Networks of marine-based data buoys (moored, profiling, and drifting) and coastal stations that measure atmospheric and oceanographic variables such as wind speed, direction, and gust; barometric pressure; air temperature; sea surface temperature; wave height and periodicity; and ocean acidification, nutrients, and dissolved oxygen. The observations from moored buoys and coastal stations are transmitted in near real-time via satellites or communication pathways to a ground receiving facility. Buoy and coastal-station observation data are used to support marine warnings and advisories and the movement of ships in and out of port and to calibrate hurricane wind speed aircraft measurements and satellite sea surface temperature, wind, and wave observations;

for directional wave measurement to study coastal erosion; and for detection of algal blooms and pathogens. These buoys impact multiple SBAs, with "high" impact in the Weather, Ocean and Coastal Resources and Ecosystems, and Transportation SBAs.

- *Survey vessels:* Survey vessels are important observation platforms for the Ocean and Coastal Resources SBA. Large-ship assets and a fleet of smaller and charter vessels form the foundation for sampling and sensing the marine ecosystems along the coast, on the continental shelves, and in the open ocean. They support the collection of living marine resources for fishery assessment and the observation and tagging of protected resources to more fully characterize the ocean system. The data, information, and results from these surveys are used widely.

5.1.4. Multi-platform Observations for Public Services and Research

The following examples of high-impact multi-platform systems and programs identified in the EOA are presented alphabetically.

- *Aquatic Resource Surveys:* The Federal Government conducts aquatic-resource surveys in partnership with State and tribal governments. These projects monitor and catalog the Nation's aquatic resources through probability-based surveys that provide nationally consistent scientific assessments of lakes, rivers, streams, coastal waters, and wetlands. These surveys track the status and vitality of aquatic resources and can be used to monitor changes in condition over time. They measure biological quality, chemical stressors, habitat stressors, and human-health indicators. These surveys have a "very high" impact in the Ecosystems SBA.
- *Forest Observations*: The Federal Government has collected and analyzed forest data for decades. Today's activities include an annual forest inventory that consists of forest-health indicators, timber-product output studies, woodland owner surveys, and a National Assessment produced every five years. These activities rely on aerial photographs, digital orthoimagery, satellite imagery, field samples, surveys, and utilization studies to project forest status, health, and coverage and assess forest-management policies and practices. They impact four SBAs, with "highest" impact in the Agriculture and Forestry and Ecosystems SBAs.

- *Water-Level Observations:* These observations monitor, measure, and assess the impact of changing water levels for government and commercial navigation, recreation, and coastal- ecosystem management. Sensors are deployed across the Nation's lakes (roughly 25% in the Great Lakes), estuaries, and ocean coastal zones. They provide the national standards for tide-and water-level reference datums used for nautical charting, coastal engineering, international treaty regulation, and boundary determination. This network of sensors impacts five SBAs, with "highest" impact in the Reference Measurements SBA and "very high" impact in the Transportation SBA. In the 2012 EOA, multiple SME teams recommended additional support to improve services and research in the Disasters, Human Health, and Transportation SBAs, and the Reference Measurements societal theme.

5.2. Experimental Air borne, Terrestrial, and Marine Observations

Federal agencies routinely conduct experimental airborne, terrestrial, and marine observations to advance human knowledge through basic and applied research, to explore technical innovation, and to improve public services. Multiple SME teams involved in the 2012 EOA recommended, when feasible and within approved agency budgets, pursuing experimental research and testing to develop additional airborne capabilities for technologies such as LIDAR. Specific recommendations included developing three-dimensional imaging for ecosystem structure, improving instruments to enable photon counting, and improving near-coastal topographic/bathymetric measurements. Multiple SME teams recommended the use of unmanned vehicles to expand coverage and efficiency of observations in areas such as land- surface, coastal, and ocean monitoring and research; species surveys; and hurricane forecasting. Several SME teams recommended additional research to improve radar system scanning rates and coverage. Finally, multiple SME teams recommended continued support for ecological observations to support fundamental research and to test emerging observation technologies and experimental techniques.

6. Agency Roles and Responsibilities for Civil Earth Observations from Space

The NASA Administrator, the Secretary of Commerce through the NOAA Administrator, and the Secretary of the Interior through the USGS Director, have responsibility for managing the U.S. Government's space- based civil Earth-observation systems. Following guidance provided by the 2010 National Space Policy[29] and in accordance with existing legal authorities, these agencies will continue the development of satellites to observe and study the Earth system, with a focus on measurements and observation programs necessary for weather forecasting, environmental monitoring, disaster-risk reduction, water-resources assessment, and climate-change research.

Observations from space are also vital to fulfilling Federal objectives across multiple SBAs. They provide critical information on atmospheric, oceanic, and terrestrial phenomena at local, regional, continental, and global scales. These observations often cover broad areas, over long periods, and with frequent revisit rates. Space provides a unique vantage point for observations, and many global data sets collected from space cannot be easily replaced by other means. Continuous high-quality observations from space are also essential for an adequate understanding of the state of the Earth system, especially on a global basis. Space-based observations support the provision of sustained public services and scientific research, and they provide a vehicle for experimental studies that advance the state of science and improve the ability to monitor and measure the Earth system.[30]

Specifically, Federal agencies will conduct sustained satellite civil Earth observations as described below. All launch dates are contingent on congressional funding of the President's annual budget requests.

6.1. Sustained Satellite Observations for Public Services

The following observations from space-based systems are important to the provision of public services. While some agencies conduct these observations to support public-service products, other agencies may also conduct research efforts in these same areas. These observations are presented alphabetically.

6.1.1. Air Quality and Ozone

The Secretary of Commerce, through the NOAA Administrator, will provide air-quality data from the suite of instruments on the Joint Polar Satellite System (JPSS) satellites, the Suomi National Polar-orbiting Partnership (S-NPP) satellite, and the Geostationary Operational Environmental Satellite (GOES) series of satellites, which measure aerosols, trace gases, and meteorological data. The Secretary of Commerce, through the NOAA Administrator, will also provide statutorily mandated sustained observations of total column ozone through the Ozone Mapping and Profiler Suite (OMPS) Nadir sensor currently on the S-NPP satellite and the JPSS-1 and -2 satellites. The NASA Administrator will study options and explore working with the Secretary of Commerce, through the NOAA Administrator, to continue ozone-profile measurements currently being made by the OMPS Limb sensor on the S-NPP satellite.

6.1.2. Land-Imaging

The NASA Administrator, together with the Secretary of the Interior through the Director of USGS, will implement a 25-year program of sustained land-imaging for routine monitoring of land-cover characteristics, naturally occurring and human-induced land-cover change, and water resources, among other uses.[31]

They will also ensure that future land-imaging data will be fully compatible with the 42-year record of Landsat observations. The NASA Administrator will be responsible for satellite development, launch, and commissioning, and the Secretary of the Interior, through the USGS Director, will be responsible for representing users' requirements; development and operation of the ground system; operational control of satellites once on orbit; and processing, archiving, and distributing land-imaging data and routine information products. The NASA Administrator and the Secretary of the Interior, through the USGS Director, will continue to collaborate to address common needs for data continuity and new technology deployment.

6.1.3. Ocean-Color Observations

The Secretary of Commerce, through the NOAA Administrator and in collaboration with the NASA Administrator, will conduct sustained ocean-color observations for marine ecosystem monitoring. The Visible Infrared Imaging Radiometer Suite (VIIRS) instrument on the S-NPP satellite accomplishes this monitoring, which will be continued with the launches of the JPSS satellite series.[32]

6.1.4. Ocean Surface and Water-Level Monitoring

The Secretary of Commerce, through the NOAA Administrator and in collaboration with the NASA Administrator and international partners, will conduct sustained observations for ocean surface and inland water-level monitoring for sea level, navigation, ocean and coastal products, and geophysical reference. The altimeter on the Jason series of satellites accomplishes this monitoring, a task that will be continued with the launch of Jason-3 scheduled in 2015.

The Secretary of Commerce, through the NOAA Administrator, will cooperate with interagency and international partners to provide active (e.g., C-band Synthetic Aperture Radar or SAR) and passive (e.g., high-resolution microwave) data for ocean-surface monitoring. The primary objectives of this monitoring will be to support safety of navigation through routine generation of ice charts and analyses, to support environmental assessment through detection and analysis of pollution (e.g., oil spills) and detection and analysis of coastal change and ocean fronts. For these applications, the Secretary of Commerce, through the NOAA Administrator, will seek to use data from the U.S. Defense Meteorological Satellite Program Special Sensor Microwave Imager instrument series, the Canadian Space Agency Radarsat-2 and Radarsat Constellation satellites, and the European Space Agency Sentinel-1 series of C-band SAR instruments.

6.1.5. Ocean-Surface Vector Winds

The Secretary of Commerce, through the NOAA Administrator and in collaboration with the NASA Administrator, will continue to cooperate with foreign partners to obtain scatterometry data for the measurement of near-ocean-surface wind speed and direction. These data sets can enhance modeling of the atmosphere, surface waves, and ocean circulation, with the potential to support a wide range of marine operations and enhance climate research and marine weather forecasting. The United States will obtain these data through continued U.S. access to data from the Advanced Scatterometer instrument on the European MetOp series of satellites and from the Oceansat-2 Scatterometer instrument on the Indian Space Research Organization satellites.

6.1.6. Space Weather Monitoring

The Secretary of Commerce, through the NOAA Administrator and in consultation with the NASA Administrator and interagency and international partners, will conduct sustained observations for space weather monitoring and prediction, which require constant operations. Specifically, the Secretary of

Commerce, through the NOAA Administrator, will provide observations of solar wind (including coronal mass ejection), solar flares, and energetic particles, and will provide radio occultation and related measurements to forecast space weather events. The Secretary of Commerce, through the NOAA Administrator, will provide these measurements through the GOES series and the Deep Space Climate Observatory (DSCOVR) satellite (to be launched in 2015) and will study options and explore working with international and interagency partners to provide these measurements beyond the design life of the DSCOVR mission.

6.1.7. Weather, Hazards, and Seasonal/Inter-annual Climate Variability

The Secretary of Commerce, through the NOAA Administrator and in collaboration with the NASA Administrator and international partners, will provide sustained satellite observations for monitoring and predicting weather and related hazards, which require constant operations. These observations comprise atmospheric sounding and imaging of the Earth; space-based environmental data relay for weather; and other products critical to the protection of lives, property, air quality, and public health. The Secretary of Commerce, through the NOAA Administrator, will also process long time-series data from these observations for reanalysis and modeling to better understand seasonal to decadal climate trends.

The Secretary of Commerce, through the NOAA Administrator, will conduct these observations through the GOES series, the JPSS program (including S-NPP and future JPSS satellites), and the Polar-orbiting Operational Environmental Satellite (POES) series. Federal agencies will collaborate to launch the next satellite in the GOES series in 2016 and the first of the future JPSS satellites in 2017.

Both NOAA's GOES and JPSS programs will experience overlapping technology infusion and the use of next-generation launch systems during the next 10-year period. Therefore, the Secretary of Commerce, through the NOAA Administrator, will take steps to re-phase future development and operational life cycles of these two programs to reduce pressure from simultaneous budget peaks in both. The Secretary of Commerce, through the NOAA Administrator, will also seek to employ related program adjustments, as appropriate, to improve flexibility and efficiency as part of a portfolio-managed approach, maintain sustained observations in the most cost-effective manner, and achieve a robust architecture for observations.

The Secretary of Commerce, through the NOAA Administrator and in collaboration with interagency and international partners, will also continue to develop and acquire GPS radio-occultation measurements to enhance weather observation and prediction through the Constellation Observing System for Meteorology, Ionosphere, and Climate (COSMIC) and COSMIC-2 missions.

6.2. Sustained Satellite Observations for Earth System Research

The NASA Administrator will conduct sustained satellite observations for research to advance the understanding of changes to the Earth system and related climate change. The Secretary of Commerce, through the NOAA Administrator and in collaboration with the NASA Administrator and other agencies, will also provide sustained observations for research on seasonal and inter-annual climate trends. Specific satellites will provide observations of the environmental phenomena described in the following subsections. These observations are presented alphabetically.

6.2.1. Aerosols and Trace Gases

In addition to the aerosols and trace gas data provided by NOAA, the NASA Administrator will provide long-term measurements of the vertical structure of aerosols, ozone, water vapor, and other important trace gases in the upper troposphere and stratosphere through the latest Stratospheric Aerosol and Gas Experiment (SAGE III) on the International Space Station (ISS). The NASA Administrator will launch SAGE- III on ISS in 2015, advancing the measurements provided by the previous Stratospheric Aerosol Measurement (SAM I and II), the SAGE I and II instruments, and the SAGE III Meteor-3M. Furthermore, the NASA Administrator will study options for continuing ozone-profile measurements planned for JPSS and explore collaboration with the Secretary of Commerce, through the NOAA Administrator, on OMPS measurements.

6.2.2. Atmospheric Carbon Dioxide

The NASA Administrator will provide global measurements of atmospheric carbon dioxide (CO_2) through the Orbital Carbon Observatory (OCO), launched in July 2014. OCO measurements will be combined with data from a ground-based network to provide information needed to better understand the processes that regulate atmospheric CO_2 and its role in the Earth's carbon cycle. Additionally, the NASA Administrator will explore using data from carbon dioxide monitoring missions planned by international partners.

6.2.3. Groundwater

The NASA Administrator will conduct precision measurements of the Earth's gravitational field that support groundwater measurements. Measurements taken through the Gravity Recovery and Climate Experiment (GRACE) help characterize the movement of underground water reservoirs and their seasonal variability. The NASA Administrator will launch the GRACE follow-on in 2017, which will continue the measurement record established by the first GRACE launched in 2002. In addition, NASA and the Indian Space Research Organisation (ISRO) are cooperating to develop the NASA-ISRO Synthetic Aperture Radar (NI-SAR) mission, an L- and S-band radar satellite that will be able to measure land subsidence in relation to groundwater resources.

6.2.4. Net Energy Balance

To understand incoming and outgoing radiant energy of the Earth system, Federal agencies have collaborated to create a 30-year-plus record of total solar irradiance and a 10-year-plus record of clouds and the Earth's radiation balance. The Secretary of Commerce, through the NOAA Administrator, and the NASA Administrator will continue these measurements as follows:

- Radiation Budget: The Secretary of Commerce, through the NOAA Administrator, will launch a new Clouds and the Earth's Radiant Energy System (CERES) sensor on JPSS-1 in 2017, and the NASA Administrator will process the sensor's data. These observations will provide continuity for the series of measurements currently produced by S-NPP and NASA's Aqua and Terra satellites. Both the Secretary of Commerce, through the NOAA Administrator, and the NASA Administrator will explore options for a future Radiation Budget Instrument.
- Total Solar Irradiance: The Secretary of Commerce, through the NOAA Administrator, will complete and launch the Total Solar Irradiance Sensor (TSIS-1). The NASA Administrator will develop plans for continuing solar irradiance observations beyond the life cycle of TSIS-1. These programs will continue the data record currently produced by NASA's Solar Radiation and Climate Experiment (SORCE) and Active Cavity Radiometer Irradiance Monitor (ACRIMSAT) satellites, as well as the Total solar irradiance Calibration Transfer Experiment (TCTE), a joint program in which NOAA and NASA collaborated to include a Total Irradiance

Measurement (TIM) sensor as a hosted payload on an Air Force STPSat-3 satellite launched in 2013.

6.3. Experimental Satellite Observations

In addition to the roles listed above, the NASA Administrator, in collaboration with other agencies, will conduct experimental observations of the Earth from space to advance human knowledge of the Earth as an integrated system. The NASA Administrator will accomplish these observations through the NASA Earth Systematic Missions and Pathfinder programs. The NASA Administrator will also continue its Venture Class program for innovative new research satellites. Through these programs, the NASA Administrator plans to launch experimental observations for the following measurements relevant to the understanding of climate and related global change (listed chronologically by actual or projected launch date):

- Global precipitation through the Global Precipitation Measurement (GPM) satellite, launched in February of 2014.
- Soil moisture through the Soil Moisture Active Passive (SMAP) satellite, to be launched in 2014.
- Cyclone generation through the Cyclone Global Navigation Satellite System (CYGNSS), to be launched in 2016.
- Ice-sheet mass balance, clouds and aerosols, and land elevation through the second Ice, Cloud, and land Elevation Satellite (ICESat-II), to be launched in 2017.
- Tropospheric pollution through the Tropospheric Emissions: Monitoring of Pollution (TEMPO) satellite, to be launched in 2019.
- Surface-water and ocean topography through the Surface Water Ocean Topography (SWOT) satellite, to be launched in 2020.
- Solid-Earth deformation, ice masses, and ecosystems through the NI-SAR satellite, to be jointly developed by NASA and ISRO.

In addition, the NASA Administrator will continue studying the feasibility of new satellite systems for observations of clouds and aerosols, land-surface characterization and deformation, hurricane formation, ecosystems classification, vegetation analysis, space-weather monitoring, and disaster-risk reduction, among others.

7. Summary Guidelines in Implementation and Coordination of the National Plan

Implementation and coordination of activities outlined in this National Plan will be conducted through Federal agencies' existing budgets, program planning, and NSTC-sponsored interagency processes. Implementation schedules, progress reviews, and funding profiles will be determined by Federal agencies in consultation with the Executive Office of the President.

The primary forum for interagency discussion and coordination of Earth observation, related data management, and related international issues is the USGEO Subcommittee of the NSTC. Within this forum, specialized working groups will facilitate triennial Earth observation assessments, enhanced data discoverability, accessibility, and usability initiatives, and U.S. participation in the intergovernmental GEO. Federal Earth observation agencies will designate appropriate representatives to the USGEO Subcommittee, and corresponding SMEs to its working groups as appropriate. To maximize the utility of Earth observations to the widest range of internal and external stakeholders and inform Federal Earth-observation activities with valuable stakeholder perspectives, USGEO Subcommittee and Working Group representatives will consult as appropriate and exchange information and expertise with CENRS subcommittees and working groups.33 Other NSTC entities that will benefit from coordination with USGEO include the NSTC Committee on Technology's Subcommittee on Network and Information Technology Research and Development and the NSTC Committee on Homeland and National Security's Topics of International Science and Technology Innovation Subcommittee. OSTP, in consultation with the USGEO Subcommittee and CENRS member agencies, will review and update this National Plan on a three-year cycle. As part of this process, OSTP will solicit and consider the input of external stakeholders and the general public.

Annex I. 2012 EOA Results

This annex provides results for the 145 high-impact observation systems identified from the 362 observation systems assessed by the 13 SBA teams of approximately 300 Federal subject-matter experts. These 145 observation systems are listed in two tiers in the tables below. Impact is indicated with respect to each of the 13 societal themes (12 SBAs and reference measurements), as described in Section 2.2.

Table 1. Tier 1 High-Impact Observation Systems (Ranked Order)

Observation System (Ranked Order)	Agency	Ag&Frst	BioDiv	Climate	Disasters	Ecosys	Energy	HumanHlth	Ocn&Cstl	Space Wx	Trans	WaterRes	Wx	RefMeas
1. Global Positioning System (GPS) satellites	DOD/USAF													
2. Next Generation Weather Radar (NEXRAD)	DOC/NOAA								*					
3. Landsat satellite	DOI/USGS, NASA										*			
4. Geostationary Operational Environmental Satellite System (GOES-NOP)	DOC/NOAA			*		*								
5. National Agriculture Imagery Program (NAIP)	USDA/FSA													
6. Airborne LIDAR	DOC/NOAA, DOD/USACE, DOI/USGS, NSF													
7. Forest Inventory and Analysis (FIA)	USDA/USFS							*						

Impact: Contributes * | Moderate | High | Very High | Highest

Table 1. (Continued)

Observation System (Ranked Order)	Agency	Ag&Frst	BioDiv	Climate	Disasters	Ecosys	Energy	HumanHlth	Ocn&Cstl	Space Wx	Trans	WaterRes	Wx	RefMeas
8. Aircraft Meteorological Observations (e.g., MDCRS, AMDAR)	DOC/NOAA, Non-USG													
9. National Water Level Observation Network (NWLON)	DOC/NOAA											*		
10. Terra satellite	NASA										*			*
11. MetOp - Polar Orbiting Operational Meteorology (satellite, EUMETSAT)[a]	Non-USG					*				*				
12. Radiosonde Observations by National Weather Service (RAOBS)	DOC/NOAA					*			*					
13. USGS Stream Gage Network	DOI/USGS						*							
14. Suomi National Polar-orbiting Partnership (S-NPP) satellite[b]	NASA, NOAA													
15. Jason satellite	DOC/NOAA, NASA	*	*											

Impact: Contributes * | Moderate | High | Very High | Highest

Note: The EOA value chain began by identifying high-impact data sets and information products, and then observing systems that generate or contribute to those data sets and information products. As a result, the EOA captured the impact of certain non-USG sources of data, including from international, non-governmental, and commercial partners. Furthermore, the EOA results are derived from the findings of the SBA subject matter experts, and may not reflect all uses of well-known observing systems. In the table above, an asterisk indicates that the EOA scored this system as contributing to the SBA but not at a moderate or high level.

[a] Both the MetOp satellites and POES system are part of the constellation required to meet the key objectives in relevant SBAs.

[b] At the time of the EOA, S-NPP had only recently been launched and its impact on key objectives in relevant SBAs had yet to be fully realized.

Table 2. Tier 2 High-Impact Observation Systems (Alphabetical Order)

Observation System (Alphabetical Order)	Agency	Ag&Frst	BioDiv	Climate	Disasters	Ecosys	Energy	HumanHlth	Ocn&Cstl	Space Wx	Trans	WaterRes	Wx	RefMeas
Advanced Composition Explorer (ACE) satellite	NASA													
Advanced National Seismic System (ANSS)	DOI/USGS													
Aerial Observers - Aerial Detection Surveys (ADS)	DOC/NOAA, USDA/USFS													
Agricultural Attaché Reports	USDA/FAS													
Agricultural Economic Data	USDA													
Airborne Geophysical Measurements	DOI/USGS, EPA													
Airborne High-Resolution Optical Imagery	DOC/NOAA, DOD, NASA, USDA													
Airborne Hyperspectral Imagery	NASA				*									

Impact: Contributes * | Moderate | High | Very High | Highest

Observation System (Alphabetical Order)	Agency	Ag&Frst	BioDiv	Climate	Disasters	Ecosys	Energy	HumanHlth	Ocn&Cstl	Space Wx	Trans	WaterRes	Wx	RefMeas
Airborne Interferometric Synthetic Aperture Radar (InSAR)	Non-USG						*							
Airborne Radar - U.S. Coast Guard	DHS/USCG													
Airport Surveillance Radar (ASR)	DOT/FAA													
American Academy of Allergy Asthma & Immunology National Allergy Bureau (AAAAI-NAB) pollen counts	Non-USG													
AmeriFlux Network	DOE													
Aqua satellite	NASA										*			
Argo Oceanographic Profilers	DOC/NOAA		*		*							*		
Atmospheric Integrated Research Monitoring Network (AIRMoN)	DOC/NOAA													

Impact: Contributes * | Moderate | High | Very High | Highest

Table 2. (Continued)

Observation System (Alphabetical Order)	Agency	Ag&Frst	BioDiv	Climate	Disasters	Ecosys	Energy	HumanHlth	Ocn&Cstl	Space Wx	Trans	WaterRes	Wx	RefMeas
Aura satellite	NASA						*							
Automated Surface Observing System (ASOS)	DOC/NOAA, DOT/FAA		*	*		*								
Automated Weather Observing System (AWOS)	DOT/FAA		*	*		*								
Autonomous Underwater Vehicles (AUVs) - temperature, salinity, and currents measurements	DOC/NOAA						*							
Breeding Bird Surveys (BBS)	DOI/FWS													
BLM Rapid Eco-regional Assessments (REA) & other federal/state agency/NGO Landscape Assessments (RA)	DOI/BLM													
Centers for Disease Control and Prevention (CDC) Surveillance	HHS/CDC				*									
Chartered Survey Vessels for Benthic Habitat observations	DOC/NOAA						*	*						

Impact: Contributes * | Moderate | High | Very High | Highest

Observation System (Alphabetical Order)	Agency	Ag&Frst	BioDiv	Climate	Disasters	Ecosys	Energy	HumanHlth	Ocn&Cstl	Space Wx	Trans	WaterRes	Wx	RefMeas
Chartered Vessels for Fish Surveys	DOC/NOAA													
Clean Air Status and Trends Network (CASTNET)	EPA													
Coastal Buoys (National Data Buoy Center Buoys, IOOS)	DOC/NOAA						*					*		
Commercial Fishery Catch Monitoring	DOC/NOAA													
Commercial High-Resolution Satellite Imagery	Non-USG													
Commercial Radar Satellites (e.g., RadarSat, TerraSAR-X)	Non-USG													
Deep-ocean Assessment and Reporting of Tsunamis (DART) Buoys	DOC/NOAA													
Defense Meteorological Satellite Program (DMSP)	DOD/USAF								*		*			

Impact: * Contributes | Moderate | High | Very High | Highest

Table 2. (Continued)

Observation System (Alphabetical Order)	Agency	Ag&Frst	BioDiv	Climate	Disasters	Ecosys	Energy	HumanHlth	Ocn&Cstl	Space Wx	Trans	WaterRes	Wx	RefMeas
Direct Reports from Farmers (Crop Reports, FSA Form 578)	USDA/FSA													
Disaster Incident Reports	EPA, DHS/FEMA													
Disaster Monitoring Constellation (DMC) satellites (foreign)	Non-USG					*								
DOE Atmospheric Radiation Measurement (ARM) Climate Research Facility	DOE							*						
Drifting Buoy Network	DOC/NOAA		*				*	*				*		
Earth Observing Satellite - 1 (EO-1)	NASA													
EPA Emissions Source Testing/ Measurement Programs (EST/M)	EPA													
EPA Water Quality Samples	EPA				*									

Impact: Contributes * | Moderate | High | Very High | Highest

Observation System (Alphabetical Order)	Agency	Ag&Frst	BioDiv	Climate	Disasters	Ecosys	Energy	HumanHlth	Ocn&Cstl	Space Wx	Trans	WaterRes	Wx	RefMeas
Farm Operator Surveys	USDA/NASS					*								
Farm Service Agency Reporting	USDA					*								
Fish and Wildlife Service Inventory and Monitoring (I&M) Program	DOI/FWS													
Fisheries Observers on Commercial Vessels	DOC/NOAA													
Global Avian Influenza Surveillance	HHS/CDC													
Global Biodiversity Information Facility Network (GBIF)	DOI/USGS	*												
Global Change Observation Mission - Water (GCOM-W) Advanced Microwave Scanning Radiometer 2 (AMSR2)	Non-USG													
Global Climate Observing System (GCOS) Upper Air Network (GUAN)	Non-USG					*					*			

Impact: Contributes * | Moderate | High | Very High | Highest

Table 2. (Continued)

Observation System (Alphabetical Order)	Agency	Ag&Frst	BioDiv	Climate	Disasters	Ecosys	Energy	HumanHlth	Ocn&Cstl	Space Wx	Trans	WaterRes	Wx	RefMeas
Global Lake Ecological Observatory Network (GLEON)	NSF													
Global Navigation Satellite System Meteorology (GNSS-Met)	DOC/NOAA			*				*			*	*		
Global Ocean Observing System (GOOS) Tropical Moorings (TAO/RAMA/PIRATA)	DOC/NOAA		*		*									
Global Sea Level Observing System (GLOSS) (international)	Non-USG										*			
Global Seismographic Network (GSN)	DOI/USGS, NSF													*
Gravity Field and Steady-State Ocean Circulation Explorer (GOCE)	Non-USG										*	*		
Gravity for the Redefinition of the American Vertical Datum (GRAV-D)	DOC/NOAA											*		
Gravity Recovery and Climate Experiment (GRACE) satellite	NASA												*	

Impact: Contributes * | Moderate | High | Very High | Highest

Observation System (Alphabetical Order)	Agency	Ag&Frst	BioDiv	Climate	Disasters	Ecosys	Energy	HumanHlth	Ocn&Cstl	Space Wx	Trans	WaterRes	Wx	RefMeas
High Frequency (HF) Coastal Radars	DOC/NOAA						*						*	
International Doppler Radar	Non-USG													
International Global Navigation Satellite Systems (GNSS)	Non-USG				*						*			
International Magnetometers	Non-USG													
Jet Propulsion Lab MODIS/ASTER airborne simulator (JPL MASTER)	NASA													
Lightning Data Buy	DOC/NOAA											*		
Long-Term Ecological & Agroecosystem Research Networks & International LTER (LTER/LTAR/ILTER)	NSF, USDA						*							
Low Level Windshear Alert System (LLWAS)	DOT/FAA													

Impact: Contributes * | Moderate | High | Very High | Highest

Table 2. (Continued)

Observation System (Alphabetical Order)	Agency	Ag&Frst	BioDiv	Climate	Disasters	Ecosys	Energy	HumanHlth	Ocn&Cstl	Space Wx	Trans	WaterRes	Wx	RefMeas
Meteosat satellite (EUMETSAT)	Non-USG		*	*				*	*					
Micro-Pulse LIDAR Network (MPLNET)	NASA						*							
Multifunctional Transport Satellite (MTSAT) Imager for Weather and Aviation Control (Japan)	Non-USG			*			*	*			*			
National Air Toxics and Trends Stations (NATTS)	EPA													
National Animal Health Surveillance System (NAHSS)	USDA/APHIS													
National Aquatic Resource Surveys (NARS)	EPA													
National Ballast Information Network	DHS/USCG													
National Cooperative Soil Survey	USDA/NRCS													

Impact: * Contributes; Moderate; High; Very High; Highest

Observation System (Alphabetical Order)	Agency	Ag&Frst	BioDiv	Climate	Disasters	Ecosys	Energy	HumanHlth	Ocn&Cstl	Space Wx	Trans	WaterRes	Wx	RefMeas
National Park Service Inventory and Monitoring (I&M)	DOI/NPS													
National Phenology Network (NPN)	DOI/NPS, DOI/USGS, NSF													
National Weather Service (NWS) Cooperative Observer Program (COOP)	DOC/NOAA													
National Weather Service Skywarn Spotters	DOC/NOAA										*		*	
NOAA and Chartered Aircraft for Protected Species Surveys	DOC/NOAA													
NOAA Global Monitoring Division Observatories	DOC/NOAA													
NOAA Profiler Network	DOC/NOAA		*	*								*		
NOAA Recreational Fish Surveys	DOC/NOAA													

Impact: Contributes * Moderate High Very High Highest

Table 2. (Continued)

Observation System (Alphabetical Order)	Agency	Ag&Frst	BioDiv	Climate	Disasters	Ecosys	Energy	HumanHlth	Ocn&Cstl	Space Wx	Trans	WaterRes	Wx	RefMeas
NOAA Ship and Chartered Ship for Coral Reef Field Surveys	DOC/NOAA													
NOAA Ships - Fish Surveys	DOC/NOAA													
NOAA Ships - Hydrographic Surveys	DOC/NOAA				*						*			
NOAA Ships - Protected Species Observations	DOC/NOAA													
Non-Federal Biological Surveys	Non-USG	*												
Ocean Reference Station Buoys	DOC/NOAA							*						
Physical Oceanographic Real-Time System (PORTS)	DOC/NOAA				*		*							
Pilot Reports (PIREPS)/Aircraft Reports (AIREPS)	DOT/FAA		*									*		

Impact: * Contributes; Moderate; High; Very High; Highest

Observation System (Alphabetical Order)	Agency	Ag&Frst	BioDiv	Climate	Disasters	Ecosys	Energy	HumanHlth	Ocn&Cstl	Space Wx	Trans	WaterRes	Wx	RefMeas
Polar-orbiting Operational Environmental Satellite System (POES)[a]	DOC/NOAA		*											
Port Radars	DHS/USCG													
Portable Water Gages	DOC/NOAA, DOI/USGS										*			
Rain Gage Networks (state, local)	Non-USG				*						*			
Regional and State Mesonetworks	DOC/NOAA		*			*					*			
Road Weather Information Systems (RWIS)	DOT/FHWA													
Satellite Pour l'Observation de la Terre (SPOT) (France)	Non-USG				*			*						
Sentinel Animals (state and local)	Non-USG						*							

Impact: * Contributes; Moderate; High; Very High; Highest

Table 2. (Continued)

Observation System (Alphabetical Order)	Agency	Ag&Frst	BioDiv	Climate	Disasters	Ecosys	Energy	HumanHlth	Ocn&Cstl	Space Wx	Trans	WaterRes	Wx	RefMeas
Shipboard Radars	DHS/USCG, DOC/NOAA, DOD/USN													
Ships of Opportunity Program (SOOP)	DOC/NOAA						*							
Small Boat Charters - Harmful Algal Bloom Water Sampling	DOC/NOAA													
Smithsonian Institution Global Earth Observatory (SIGEO)	SI													
SNOwpack TELemetry (SNOTEL)	USDA/NRCS						*							
Soil Climate Analysis Network (SCAN)	USDA/NRCS						*							
Soil Moisture and Ocean Salinity Mission (SMOS) satellite (ESA)	Non-USG													
Solar and Heliospheric Observatory (SOHO) satellite	NASA													

Impact: Contributes * | Moderate | High | Very High | Highest

Observation System (Alphabetical Order)	Agency	Ag&Frst	BioDiv	Climate	Disasters	Ecosys	Energy	HumanHlth	Ocn&Cstl	Space Wx	Trans	WaterRes	Wx	RefMeas
Solar Dynamics Observatory (SDO) satellite	NASA													
Solar Electro-Optical Network (SEON)	DOD/USAF													
Solar Terrestrial Relations Observatory (STEREO) satellites	NASA													
Sound Velocimeter for hydrographic surveys	DOC/NOAA													
Spatially Enhanced Broadband Array Spectrograph System (SEBASS)	NASA													
State and Local Stream Gage Networks	DOI/USGS	*			*		*	*			*			*
State and Local Water Level Networks	DOI/USGS													
Terminal Doppler Weather Radar (TDWR)	DOT/FAA							*					*	

Impact: Contributes * | Moderate | High | Very High | Highest

Table 2. (Continued)

Observation System (Alphabetical Order)	Agency	Ag&Frst	BioDiv	Climate	Disasters	Ecosys	Energy	HumanHlth	Ocn&Cstl	Space Wx	Trans	WaterRes	Wx	RefMeas
Toxics Release Inventory (TRI)	EPA													
Tropical Rainfall Measuring Mission (TRMM) satellite	NASA		*		*				*					
U.S. Army Corps of Engineers Hydrographic Surveys	DOD/USACE													
U.S. Census Data on Population and Roads	DOC/ Census													
Unmanned Aircraft Systems (UAS) - Surveys and Surveillance	DOC/NOAA, DOD, DOI/USGS				*				*					
US Gap Analysis Program (GAP) - assessing biodiversity conservation	DOI/USGS													
USGS Geomagnetic Observatories	DOI/USGS													*
USGS In Situ Water Quality Sensors	DOI/USGS	*	*		*						*			

Impact: Contributes * Moderate High Very High Highest

Observation System (Alphabetical Order)	Agency	Ag&Frst	BioDiv	Climate	Disasters	Ecosys	Energy	HumanHlth	Ocn&Cstl	Space Wx	Trans	WaterRes	Wx	RefMeas
USGS Inner Shelf Vessels for Benthic Habitat Observation	DOI/USGS													
USGS National Groundwater Stations	DOI/USGS				*		*	*						
USGS Water Quality Samples	DOI/USGS				*									
Very Long Baseline Interferometry (VLBI)	NSF													
Voluntary Observing Ships (VOS)	DOC/NOAA		*					*				*		
Water Resources Field Experiments	DOI/USGS													
WC-130 - U.S. Air Force - Hurricane Hunter aircraft	DOD/USAF		*						*		*	*		
WMO Global Observing System (GOS) - Surface	Non-USG				*	*					*			

Impact: * Contributes; Moderate; High; Very High; Highest

Table 2. (Continued)

Observation System (Alphabetical Order)	Agency	Ag&Frst	BioDiv	Climate	Disasters	Ecosys	Energy	HumanHlth	Ocn&Cstl	Space Wx	Trans	WaterRes	Wx	RefMeas
WMO Global Observing System (GOS) - Upper Air	Non-USG			*		*								
World Aviation Forecast System Internet File Services (WIFS)	DOT/FAA													

Impact: Contributes * | Moderate | High | Very High | Highest

[a] Both MetOp satellites and POES system are part of the constellation required to meet the key objectives in relevant SBAs.

Note: The EOA value chain began by identifying high-impact data sets and information products, and then observing systems that generate or contribute to those data sets and information products. As a result, the EOA captured the impact of certain non-USG sources of data, including from international, non-governmental, and commercial partners. Furthermore, the EOA results are derived from the findings of the SBA subject matter experts, and may not reflect all uses of well-known observing systems. In the table above, an asterisk indicates that the EOA scored this system as contributing to the SBA but not at a moderate or high level.

ANNEX II. CAVEATS FOR UNDERSTANDING AND INTERPRETING THE 2012 EARTH OBSERVATION ASSESSMENT

1. *Comprehensiveness.* The list of systems assessed in the EOA and identified in Annex I of this National Plan is not a comprehensive inventory of all systems, but rather an analysis of systems with significant impact as identified by a broad range of subject-matter experts. In addition, the EOA results include only the current portfolio of deployed systems. Planned and future systems (JPSS, NEON, etc.) were not analyzed. Finally, the EOA does not encompass the full range of possible objectives within each SBA. Therefore, certain agency mission objectives and systems important to those objectives may not be adequately reflected.
2. *Significance of All Systems.* All systems identified in the EOA have significant impact on key objectives identified by expert teams under each SBA.
3. *Assessment Results and the Budget Process: Impact vs. Value.* The EOA did not include cost data.

   Therefore, the EOA results convey **the impact, not the value**, of the observing systems identified by expert teams. The Assessment and this National Plan are meant to provide useful input to the budget-review process and to complement (but not substitute for) information and methods traditionally used in these reviews.
4. *Considering the Enterprise as a Whole.* The list of high-impact observing systems in Annex I is best used as a device for understanding the relative impact of each system on the broader, national Earth observation enterprise. Decisions about observing systems in one agency can have a dramatic effect (positive or negative) on the ability of other agencies to perform their Earth-science missions. Earth- observation investment decisions and adjustments must therefore be made through a coordinated process to avoid unintended consequences on the entire enterprise. For example, the EOA results alone should not be used as the basis for eliminating individual systems. Such a rough-cut use of this system would have devastating effects on specific areas of Earth science and services delivered to taxpayers.

5. *Even Weighting of the SBAs*. The work of the EOA is organized around evenly weighted "Societal Benefit Areas," because an economic analysis of the value of the information to society is well beyond the scope of the EOA. Therefore, the list shows relative, rather than absolute priority, across all areas. It is appropriate to review the EOA results not just in the larger list, but by SBA to see how the relative impact of a given system rises or falls, depending on the area of emphasis.
6. *Multiple- and Special-Purpose Systems*. Reviewers of these results should understand that systems with a high impact across multiple SBAs score highest in the EOA. Nevertheless, many significant systems that did not meet the criteria established for Tier 1 (i.e., impact across *many* SBAs) are considered essential to the *specific* SBA they uniquely serve. Many of these High-Impact Special Purpose Systems are included in as Tier 2 systems, listed in Table 2 in Annex I.
7. *A Companion to Other Analyses of Research-Observing-System Priorities*. The analysis presented here is best understood as a companion to other analyses of Earth-observation priorities that are provided to the Federal Government, most importantly the National Academies report *Earth Science and Applications from Space* (2007), known as the decadal survey. The EOA provides two new perspectives to complement the work of the decadal survey: (a) the inclusion of non-satellite systems and (b) a robust analysis of the impact of all systems on the delivery of services to society. Fundamental research about the Earth system underpins each of the 12 SBAs, and each team was invited to consider research priorities critical for their area. The EOA process, however, was fundamentally applications- oriented, and the constraints of time and the breadth of the analysis prevented a full accounting of research needs in every area. Therefore the Assessment's results for research observation systems may not reflect the full impact of those systems on climate and other research needs. Aside from this dimension, the EOA does capture the impact of current research systems on specific societal-benefit applications.
8. *International Interests*. The Nation's Earth-observing systems have important impacts on U.S. foreign- security-policy interests that were not included in this analysis. Agency commitments and agreements

with international partners on collaborative systems were also not factored into the impact analysis.

ANNEX III. ABBREVIATIONS

ACRIMSAT	Active Cavity Radiometer Irradiance Monitor
AMDAR	Aircraft Meteorological Data Relay
APHIS	Animal and Plant Health Inspection Service, Department of Agriculture
BEDI	Big Earth Data Initiative
BISON	Biodiversity Information Serving Our Nation
BLM	Bureau of Land Management, Department of the Interior
CDC	Centers for Disease Control and Prevention, Department of Health and Human Services
CENRS	Committee on Environment, Natural Resources, and Sustainability
CERES	Clouds and the Earth's Radiant Energy System
CO2	carbon dioxide
COSMIC	Constellation Observing System for Meteorology, Ionosphere, and Climate
CYGNSS	Cyclone Global Navigation Satellite System
DHS	Department of Homeland Security
DOC	Department of Commerce
DOD	Department of Defense
DOE	Department of Energy
DOI	Department of the Interior
DOT	Department of Transportation
DSCOVR	Deep Space Climate Observatory
ECV	Essential Climate Variable
EOA	Earth Observation Assessment
EPA	Environmental Protection Agency
ESA	European Space Agency
FAA	Federal Aviation Administration, Department of Transportation
FAS	Foreign Agricultural Service, Department of Agriculture
FHWA	Federal Highway Administration, Department of Transportation

FIA	Forest Inventory and Analysis, United States Forest Service, Department of Agriculture
FEMA	Federal Emergency Management Agency, Department of Homeland Security
FSA	Farm Service Agency, Department of Agriculture
FWS	Fish and Wildlife Service, Department of the Interior
FY	fiscal year
GCOS	Global Climate Observing System
GEO	Group on Earth Observations
GEOSS	Global Earth Observation System of Systems
GOES	Geostationary Operational Environmental Satellite
GPM	Global Precipitation Measurement
GPS	Global Positioning System
GRACE	Gravity Recovery and Climate Experiment
HHS	Department of Health and Human Services
ICESat	Ice, Cloud, and land Elevation Satellite
IOOS®	Integrated Ocean Observing System
ISRO	Indian Space Research Organisation
ISS	International Space Station
JPSS	Joint Polar Satellite System
LIDAR	Light Detection and Ranging
MDCRS	Meteorological Data Collection and Reporting System
NASA	National Aeronautics and Space Administration
NASS	National Agricultural Statistics Service, Department of Agriculture
NEOTF	National Earth Observation Task Force
NOAA	National Oceanic and Atmospheric Administration, Department of Commerce
NPS	National Park Service, Department of the Interior
NRCS	Natural Resources Conservation Service, Department of Agriculture
NSF	National Science Foundation
NSTC	National Science and Technology Council
OCO	Orbital Carbon Observatory
OMB	Office of Management and Budget
OMPS	Ozone Mapping and Profiler Suite
OSTP	Office of Science and Technology Policy
PACE	Preliminary Advanced Colloids Experiment
POES	Polar-orbiting Operational Environmental Satellite

RFI	Request for Information
S-NPP	Suomi National Polar-orbiting Partnership
SAGE	Stratospheric Aerosol and Gas Experiment
SAM	Stratospheric Aerosol Measurement
SAR	Synthetic Aperture Radar
SBA	societal benefit area
SI	Smithsonian Institution
SMAP	Soil Moisture Active Passive
SME	subject matter expert
SORCE	Solar Radiation and Climate Experiment
SWOT	Surface Water Ocean Topography
TCTE	Total solar irradiance Calibration Transfer Experiment
TEMPO	Tropospheric Emissions: Monitoring of Pollution
TIM	Total Irradiance Measurement sensor
TSIS	Total Solar Irradiance Sensor
USACE	United States Army Corps of Engineers
USAF	United States Air Force
USCG	United States Coast Guard
USDA	United States Department of Agriculture
USFS	United States Forest Service, Department of Agriculture
USG	United States Government
USGEO	United States Group on Earth Observations
USGS	United States Geological Survey, Department of the Interior
USN	United States Navy
VIIRS	Visible Infrared Imaging Radiometer Suite

End Notes

[1] The term "Earth observations" refers to data and products derived from Earth-observing systems and surveys. The term "observing systems" refers to one or more sensing elements that directly or indirectly collect observations of the Earth, measure environmental parameters, or survey biological or other Earth resources (land surface, biosphere, solid Earth, atmosphere, and oceans). A more detailed definition is provided in Section 2: Definitions and Context.

[2] National Earth Observations Task Force, *National Strategy for Civil Earth Observations*, Washington, DC: Office of Science and Technology Policy, April 2013, p. 24, http://www.whitehouse.gov/sites/default/files/microsites/ostp/nstc_2013_earthobsstrategy.pdf; Executive Order No. 13642 "Making Open and Machine Readable the New Default for Government Information," May 9, 2013, http://www.whitehouse.gov/the-press-office/2013/05/09/executive-order-making-open-and-machine-readable-new-default- government-.

[3] Defense and national-security requirements and considerations are not covered by this National Plan, though the use of defense and national-security assets for civil purposes is included. The Department of Defense is responsible for developing solutions for defense Earth observation requirements to support military operations and makes data available for civil agency use as appropriate. Coordination and oversight of civil agency use of national-security classified collections is performed by the interagency Civil Applications Committee. See the National Earth Observations Task Force, *National Strategy for Civil Earth Observations*, Washington, DC: Office of Science and Technology Policy, April 2013, p. 13.

[4] *Earth Observations and Global Change*, Center for Strategic and International Studies (CSIS), 2008, p. 10. http://csis.org/files/media

[5] Non-Federal entities encompass State, regional, local, and tribal governments; nongovernmental organizations; academia; citizen scientists; commercial firms; international organizations; and foreign governments.

[6] National Aeronautics and Space Administration Authorization Act of 2010 (Public Law 111–267): *SEC. 702. INTERAGENCY COLLABORATION IMPLEMENTATION APPROACH. - The Director of OSTP shall establish a mechanism to ensure greater coordination of the research, operations, and activities relating to civilian Earth observation of those Agencies, including NASA, that have active programs that either contribute directly or indirectly to these areas. This mechanism should include the development of a strategic implementation plan that is updated at least every 3 years, and includes a process for external independent advisory input. This plan should include a description of the responsibilities of the various Agency roles in Earth observations, recommended cost-sharing and procurement arrangements between Agencies and other entities, including international arrangements, and a plan for ensuring the provision of sustained, long-term space- based climate observations. The Director shall provide a report to Congress within 90 days after the date of enactment of this Act on the implementation plan for this mechanism.*

[7] National Earth Observations Task Force, *National Strategy for Civil Earth Observations*, Washington, DC: Office of Science and Technology Policy, April 2013.

[8] This National Plan contains both bulleted and numbered lists. Numbered lists indicate the order of priority, whereas bulleted lists indicate equal priority.

[9] Model outputs are generally excluded from this definition; however, some observing systems produce and record measures and observations that may require sensor models to process raw observations to a form in which they are exploitable.

[10] Group on Earth Observations, *Global Earth Observation System of Systems (GEOSS): 10-Year Implementation Plan Reference Document*, 2005, http://www.earthobservations.org/documents/10-Year%20Plan%20Reference% 20Document.pdf; Interagency Working Group on Earth Observations, *Strategic Plan for the U.S. Integrated Earth Observation System*, NSTC Committee on Environment and Natural Resources, 2005, http://www.whitehouse.gov/sites/default/files/microsites/ostp/eocstrategic_plan.pdf .

[11] See the National Strategy for a complete description of the assessment process.

[12] The EOA provided an innovative approach to understanding the impacts of Federal Earth-observation systems. The results of the EOA, however, must be considered in the context of the limitations of this assessment, which is the first of its kind. For example, fundamental research about the Earth system underpins each of the 12 SBAs, and each team was invited to consider research priorities critical for its area. The EOA process, however, was fundamentally service-oriented, and the constraints of time and the breadth of the analysis prevented a full accounting of research needs in every area. Therefore the EOA's results for research observation systems may not reflect the full impact of those systems on climate and other research needs. See Annex II for a full list of caveats.

[13] Executive Office of the President, *The President's Climate Action Plan*, 2013, http://www.whitehouse.gov/share/climate- action-plan.

[14] Office of Science and Technology Policy, *National Plan for Civil earth Observations; Request for Information*, Office of the Federal Register, 2013, https://www.federalregister.gov/articles/2013/11/12/2013-26890/national-plan-for-civil-earth- observations-request-for-information; see the USGEO web page for a list of public responses to the RFI for the development of the National Plan for Civil Earth Observations, http://www.whitehouse.gov/administration/eop/ostp/library/shareyourinput/earthobsrfi.

[15] By focusing on duration, this new framework overcomes the conceptual limitations of the traditional categories of "research" and "operational" observations, which conflate three elements of Earth observation planning: purpose, duration, and state. The purpose of the system reflects the rationale for the agency's deployment of the system—to deliver public services or to conduct research in the public interest (including basic research). Duration reflects the time period over which the agency intends to perform the observation. State refers to the status of an observing system as it evolves—from testing, through development, to operations—no matter the intended purpose or duration of the agency commitment to the observation system. Whereas research clearly refers to purpose, the term "operational" has come to mean purpose, duration, and state, particularly in the defense and aerospace communities, in the sense of "supporting ongoing operations." Under the National Plan's new framework, civil Earth observation systems previously characterized as "operational" are included under "sustained observations for public services."

[16] For the purposes of this National Plan, a commitment to maintain observations for seven years or more is considered to be the minimum threshold necessary for provision of long-term services and research in the public interest. Most experimental observation programs are not planned to exceed seven years.

[17] These measurement categories were derived by reviewing the primary purpose of the 15 highest impact systems identified by the EOA across all SBAs. They are designated as Tier 1 systems in this National Plan. The examples of questions and phenomena supported by each measurement group are meant to be illustrative and do not represent the fullest range of possible uses for these measurements. A list of systems supporting Tier 1 measurements can be found in Table 1 in Annex I.

[18] The measurement categories in Tier 2 were derived by analyzing the primary purpose of the observing systems not covered in Tier 1. Tier 2 includes the remaining high-impact systems identified in the EOA across all SBAs, and also includes additional special-purpose systems designated as high impact to specific individual SBAs. A list of systems supporting Tier 2 measurements can be found in Table 2 in Annex I.

[19] The tiers prioritize measurements, not the SBAs that the measurements support (e.g., understanding geo-hazards is not ranked below understanding weather hazards, but weather hazard measurements support a wider range of uses than geo- hazard measurements).

[20] The 2012 EOA included climate and related global change research needs as an SBA, but did not systematically prioritize research observations (see Footnote 13). Therefore, for the purposes of the National Plan, these measurement categories have been identified but not prioritized. Future assessments will address the question of priorities for sustained research in consultation with external stakeholders, including the National Academies, as appropriate.

[21] Agency processes will include consultation with external stakeholders, such as the National Academies, as appropriate.

[22] These efforts are in accordance with existing policy guidance on open data, including Executive Order No. 13642, "Making Open and Machine Readable the New Default for Government Information," and Office of Management and Budget (OMB) Memorandum M-13-13, "Open Data Policy—Managing Information as an Asset," both dated May 9, 2013.

[23] BEDI follows the life-cycle data-management principles articulated in the National Strategy and is designed to help Earth- science agencies comply more fully Executive Order No. 13642.

[24] Additionally, certain infrastructure can serve multiple purposes. For example, ships are able to place both researcher and technology at the site of the science and remain deployed "on effort" for extended periods of time.

[25] The *Federal Oceanographic Fleet Status Report* is a an example of an effort to advance the efficient and effective operation of a Federal fleet, in part for the purpose of observations at the lowest possible life-cycle costs (http://www.whitehouse.gov/sites/default/files/federal_oceanographic_fleet_status_report.pdf).

[26] To further the objectives of GEOSS, U.S. agencies participate in various international organizations and coordination groups, including: the Intergovernmental Oceanographic Commission, United Nations Framework Convention on Climate Change, United Nations International Strategy for Disaster Reduction, World Meteorological Organization, and Committee on Earth Observation Satellites, among others.

[27] Successful examples of these types of initiatives include USGS's "Did You Feel It?" program, which collects crowd-sourced data to better understand and measure earthquakes, and its Biodiversity Information Serving Our Nation (BISON) program, which aggregates species biodiversity data from a variety of sources, including museums, research studies, and citizen science programs. The President's Council of Advisors on Science and Technology has proposed the creation of an Ecoinformatics-based Open Resources and Machine Accessibility database called EcoINFORMA with the intention of making Federal data on environmental health widely accessible. Approximately 1,300 certified non-Federal, automated weather observing systems feed observation data into the Federal Aviation Administration's national automation systems. The non- Federal systems are owned and operated by various airport operating authorities including State, city, county, and local governments; municipalities; homeowners associations; and individual citizens.

[28] For climate monitoring and research, agencies may refer to guidance provided by the Global Climate Observing System (GCOS). GCOS has identified 50 Essential Climate Variables (ECVs), which are required for systematic monitoring of, and research on, the Earth's changing climate. Fundamental Climate Data Records derived from airborne, terrestrial (including freshwater), marine and satellite-based observing systems are needed to address ECV information needs.

[29] *National Space Policy of the United States of America*, June 28, 2010, pp. 12–13, http://www.whitehouse.gov/the-press- office/fact-sheet-national-space-policy.

[30] Data from NASA programs are used to improve public services for air quality, ocean color, and space weather, for example.

[31] A robust land-imaging program requires data to supplement optical imagery. Non-optical sensing capabilities such as radar, LIDAR, and gravity measurements are needed to assess natural and anthropogenic hazards and to measure changes to topography, biomass, ecosystem flux, soil moisture, coastal and inland land subsidence, surface water, groundwater, and glaciers.

[32] NOAA will also seek to use data from the European Space Agency Sentinel-3 series of SAR instruments, launching in 2015.

[33] These groups might include, but are not limited to, the Subcommittee on Global Change Research, the Subcommittee on Disaster Reduction, the Subcommittee on Ecological Services, the Interagency Arctic Research Policy Subcommittee, the Subcommittee on Water Availability and Quality, the Subcommittee on Air Quality Research, and the Subcommittee on Ocean Science and Technology.

INDEX

A

access, vii, 3, 6, 12, 14, 15, 16, 17, 18, 20, 21, 22, 45, 46, 48, 49, 50, 53, 59, 60, 61, 67, 69, 71, 75, 87, 90, 96, 97, 105
accessibility, 21, 45, 60, 63, 75, 90, 110
accounting, 132, 137
acquisitions, 22
adaptation, 3, 29, 30
advancement, 8, 82
aerosols, 30, 34, 81, 86, 104, 107, 109
aerospace, 137
agency collaboration, 77
agility, 92
agriculture, vii, 7, 26, 27, 28, 42, 73, 78, 85, 99
Air Force, 109, 135
air pollutants, 36
air quality, 30, 36, 78, 79, 80, 94, 106, 138
air temperature, 36, 86, 100
air-traffic, 40, 41, 99
algorithm, 88
alkalinity, 81
America COMPETES Act, 24, 65
animal disease(s), 27
aquaculture, 7, 74
assessment, 2, 3, 5, 8, 9, 10, 11, 12, 13, 16, 32, 39, 58, 64, 68, 69, 71, 72, 73, 75, 84, 87, 101, 103, 105, 136
assets, 6, 13, 32, 46, 48, 57, 64, 66, 101, 136
asthma, 36
atmosphere, vii, 4, 31, 38, 43, 59, 73, 97, 105, 135
authenticity, 52
authorities, 15, 97, 103, 138
automate, 89
automation, 92, 138
awareness, 36, 59, 63

B

bandwidth, 51
barometric pressure, 100
base, 89
basic research, 137
beneficiaries, 3
benefits, 2, 3, 5, 8, 9, 10, 24, 26, 31, 33, 34, 36, 40, 41, 42, 69, 78, 88
biodiversity, 7, 28, 29, 37, 73, 80, 86, 98, 138
bioenergy, 34, 35
bioinformatics, 16
biological samples, 17
biomass, 138
biosensors, 37
biosphere, 4, 73, 81, 97, 135
biotechnology, 29, 94
blueprint, vii
bottom-up, 30
Bureau of Land Management, 133

businesses, 1, 31, 55, 71

C

caching, 51
calibration, 43, 49, 86, 89, 92, 93, 94
campaigns, 82
carbon, 27, 30, 34, 81, 98, 107, 133
carbon dioxide, 30, 107, 133
casting, 3, 51
categorization, 77
CDC, 133
challenges, 13, 24, 77
chemical(s), 33, 36, 40, 41, 43, 86, 101
China, 58, 95
circulation, 105
cities, 31, 35
citizens, 3, 55
civil liberties, 6, 13
clarity, 77
classes, 80
classification, 109
Clean Air Act, 25
clean energy, 58
climate, 1, 3, 7, 12, 15, 16, 26, 27, 29, 30, 31, 32, 33, 34, 36, 37, 38, 42, 43, 45, 58, 65, 71, 74, 75, 78, 79, 80, 81, 84, 86, 97, 98, 99, 103, 105, 106, 107, 109, 132, 136, 137, 138
climate change, 16, 27, 29, 30, 31, 33, 36, 38, 45, 58, 81, 86, 98, 107
CO2, 31, 37, 107, 133
coal, 34
Coast Guard, 135
coastal communities, 38
coastal ecosystems, 38, 86
collaboration, vii, 5, 6, 55, 70, 86, 87, 88, 95, 96, 104, 105, 106, 107, 109
color, 92, 104, 138
commerce, 29, 37, 38, 84
commercial, 9, 51, 64, 70, 78, 88, 91, 94, 98, 102, 113, 130, 136
Committee on Environment and Natural Resources, 64, 65, 136
commodity, 26
commodity markets, 26
common rule, 23
communication, 22, 33, 39, 60, 100
communication technologies, 39
communities, 13, 17, 22, 23, 29, 31, 33, 37, 38, 46, 91, 137
community, vii, 5, 23, 46, 51, 54, 55, 59, 60, 76, 88
compatibility, 20, 21, 23
compensation, 28
competition, 27
competitiveness, 13
complement, 9, 57, 75, 131, 132
complexity, 1, 4, 63, 67, 71
compliance, 26, 27
composites, 28
composition, 34, 55, 80
compounds, 43
computer, 17
confidentiality, 2, 17, 46
configuration, 92, 93
Congress, 14, 24, 25, 55, 56, 58, 59, 61, 65, 68, 69, 71, 83, 136
conservation, 27, 28, 29, 32, 38
conserving, 7, 73
consolidation, 91, 96
construction, 69, 83, 86
consumers, 40
consumption, 16, 30, 42
contaminant, 41
contamination, 36, 38
continuous data, 78, 80
convergence, 23
cooperation, 64, 66, 92
coordination, 5, 6, 8, 12, 14, 18, 24, 55, 56, 58, 59, 65, 67, 68, 70, 71, 72, 91, 92, 93, 95, 110, 136, 138
copper, 34
corrosion, 32
cost, 11, 13, 23, 45, 55, 60, 61, 65, 69, 87, 88, 89, 91, 92, 93, 94, 106, 131, 136
cost saving, 45, 69, 87, 91, 92
crises, 3
critical infrastructure, 32
crop, 16, 26, 27, 28, 33

crop insurance, 26
cross-program, 91
CSA, 62
cycles, 18, 87, 92
cyclones, 31

D

data center, 51
data collection, 6, 14, 45, 50, 51, 79, 81, 99
data distribution, 23
data processing, 92
data set, 18, 19, 20, 21, 23, 35, 36, 42, 48, 53, 54, 60, 84, 86, 91, 94, 98, 103, 105, 113, 130
data structure, 52
database, 16, 138
decision makers, 1, 40, 55, 71, 95
decision-making process, 17
decomposition, 34
deficiencies, 27
deforestation, 26
deformation, 81, 109
degradation, 26, 27, 37, 39
Department of Agriculture, 16, 24, 83, 133, 134, 135
Department of Commerce, 24, 57, 133, 134
Department of Defense, 25, 133, 136
Department of Energy, 25, 83, 133
Department of Health and Human Services, 134
Department of Homeland Security, 25, 133
Department of the Interior, 25, 63, 133, 134, 135
Department of Transportation, 83, 133
Departments of Agriculture, 62, 63, 97
depth, 43
destruction, 49
detection, 33, 41, 86, 89, 98, 101, 105
DHS, 133
direct observation, 10
disaster, 1, 26, 31, 32, 33, 81, 103, 109
disaster relief, 26
discharges, 29, 38
discrimination, 46
diseases, 27, 36
disposition, 52
dissolved oxygen, 100
distribution, 28, 29, 30, 35, 36, 37, 38, 46, 60, 92
diversity, 14, 28, 29
DOC, 133
DOI, 16, 63, 133
DOT, 83, 133
draft, 14, 55
drawing, 58
drought, 28, 43, 79, 81, 100
durability, 18, 21, 52

E

early warning, 26, 31, 89
Earth observing systems, vii, 83
earthquakes, 28, 31, 32, 78, 80, 138
Earth-system data, vii, 67, 71
ecological processes, 34
economic activity, 85
economic change, 28
economic development, 34, 35, 36
economic well-being, 1, 71
ecosystem, 1, 4, 7, 24, 28, 29, 30, 33, 34, 35, 38, 55, 71, 74, 81, 85, 86, 89, 102, 104, 138
elaboration, 5
emergency, 5, 41, 42
emergency management, 41
encouragement, 24
endangered, 38
end-users, 48
energy, 7, 34, 35, 36, 41, 42, 74, 78, 81, 108
energy emission, 36
engineering, 93, 102
England, 66
entrepreneurship, 75, 97
environment, 24, 29, 32, 34, 63
environmental change, 30, 86
environmental factors, 7, 38, 74
environmental impact, 35, 40, 65
environmental management, 78

Environmental Protection Agency, 25, 97, 133
environmental threats, 55
environments, 38, 43, 93
EPA, 97, 133
equipment, 32, 93, 94
erosion, 101
European Commission, 58, 62, 95
evacuation, 32
evaporation, 41, 42, 43
evapotranspiration, 82
evolution, 5, 72, 94
Executive Order, 70, 96, 136, 137, 138
expertise, 9, 10, 58, 95, 110
exploitation, 88, 96
exposure, 32
extinction, 29
extraction, 34, 35
extreme weather events, 32

F

FAA, 133
Facebook, 54
famine, 26
farmers, 1, 26, 55, 71
farmland, 30
FAS, 133
federal agency, 55
Federal Emergency Management Agency, 134
Federal funds, 12, 45
Federal Geographic Data Committee, 16, 63
Federal Government, 1, 2, 4, 5, 7, 13, 15, 45, 71, 73, 82, 99, 101, 132
Federal Highway Administration, 133
Federal Register, 137
FEMA, 134
FGDC, 48, 63
FHWA, 133
fiber, 26, 27
financial, 95
fire detection, 78, 79
fires, 28, 32
first responders, 71
Fish and Wildlife Service, 134
fisheries, 7, 74, 80
fitness, 48
fixation, 30
flexibility, 69, 83, 106
flight, 94
flooding, 32, 79
floods, 27, 31, 78, 79, 82
food, 26, 27, 36, 41, 42, 78, 79
food security, 26, 36, 41, 42
force, 57
forecasting, 7, 26, 33, 35, 42, 74, 78, 81, 84, 86, 92, 93, 99, 102, 103, 105
foreign policy, 95
forest fire, 28
formation, 34, 109
freedom, 21
freshwater, 4, 7, 33, 42, 71, 73, 74, 80, 81, 97, 138
fuel loads, 32
funding, 13, 21, 45, 61, 64, 67, 70, 82, 83, 87, 94, 103, 110
funds, 77

G

general education, 81
genetic diversity, 28
geolocation, 8, 43, 74
Global Change Research Act, 25, 61
global climate change, 30
global scale, 27, 30, 86, 103
GOES, 104, 106, 134
goods and services, 16
governance, 56, 58
governments, 26, 58, 64, 66, 95, 101, 136
GPS, 39, 63, 79, 85, 89, 107, 134
gravitational field, 43, 108
gravity, 138
grazing, 26, 27
greenhouse, 26, 27, 29, 30, 81, 86, 96
greenhouse gas(es), 26, 27, 30, 81, 86, 96
grids, 7, 32, 39, 74
groundwater, 41, 82, 86, 98, 108, 138
growth, 28

guidance, 9, 10, 14, 18, 20, 47, 50, 56, 60, 68, 69, 72, 83, 90, 103, 137, 138
guidelines, 2, 4, 12, 20, 23, 26, 50, 52, 71, 72
Gulf of Mexico, 47

H

habitat, 34, 37, 38, 41, 101
handheld devices, 82
harmonization, 6, 50
harvesting, 37
hazards, 31, 32, 37, 38, 79, 85, 106, 137, 138
health, 1, 27, 28, 32, 33, 36, 37, 41, 55, 71, 101, 138
Health and Human Services, 133
health care, 55
health risks, 33
height, 100
HHS, 134
homeowners, 138
homes, 31
hot spots, 32
human, vii, 3, 4, 5, 7, 28, 29, 30, 31, 32, 33, 36, 37, 38, 41, 42, 49, 55, 65, 67, 68, 70, 73, 74, 76, 79, 80, 82, 85, 89, 94, 97, 101, 102, 104, 109
human activity, 3, 28
human health, 7, 29, 32, 36, 41, 74, 79
humidity, 31, 81, 86, 99
hurricanes, 28, 33
hydrosphere, 97

I

identification, 7, 74, 75, 94
image(s), 33, 65, 66, 96
imagery, 17, 92, 93, 99, 101, 138
imaging systems, 96
improvements, 40, 47, 69, 78, 82, 83, 84, 87, 93
individuals, 31, 42, 50, 54, 55, 76
industries, 34, 42
industry, 13, 67, 70, 76, 85
information technology, 77
infrastructure, 12, 29, 31, 32, 33, 40, 56, 58, 64, 70, 71, 78, 84, 88, 91, 94, 138
insects, 27
institutions, 55, 76, 100
integration, 2, 6, 12, 15, 17, 21, 56, 77, 87, 90, 91, 94
integrity, 49, 50
interagency coordination, 1, 4, 6, 21
interagency planning, vii
interface, 37
International Space Station, 107, 134
international standards, 23, 48, 52, 60, 65
interoperability, 15, 21, 22, 23, 47, 48, 50, 69, 75, 87, 89, 90, 91
investment(s), vii, 1, 3, 4, 5, 12, 24, 35, 55, 56, 57, 59, 69, 70, 82, 84, 89, 131
irrigation, 30
islands, 33
issues, 12, 22, 53, 56, 70, 110
Italy, 54

J

Japan, 96

L

lakes, 43, 101, 102
landscape(s), 27, 33, 34, 43
lead, 9, 37, 49, 54
leadership, 6, 8, 14, 58
legislation, 25
lens, 54
life cycle, 16, 17, 18, 19, 20, 21, 23, 48, 49, 54, 65, 106, 108
lifetime, 17, 39, 96
light, 85, 98
light detection and ranging, 85, 98
local government, 138
long-term services, 137
lying, 30

M

magnitude, 4, 14, 31, 49
majority, 46, 68, 79
mammal, 36
man, 38
management, vii, 2, 3, 4, 5, 6, 7, 12, 13, 14, 15, 18, 19, 20, 21, 22, 23, 27, 29, 33, 34, 35, 38, 40, 41, 48, 49, 50, 53, 54, 56, 69, 70, 71, 74, 75, 77, 79, 80, 85, 87, 88, 89, 90, 91, 95, 99, 101, 102, 110, 138
mapping, 79, 89, 98
marine fish, 38
mass, 30, 80, 106, 109
materials, 35, 40, 82
matter, 74, 80, 110, 113, 130, 131, 135, 137
media, 17, 47, 50, 60, 136
melting, 30
membership, 57
metabolism, 35
military, 99, 136
mineral resources, 7, 34, 74
mission(s), vii, 6, 13, 21, 25, 39, 43, 45, 47, 56, 60, 62, 65, 67, 70, 72, 77, 84, 95, 106, 107, 108, 131
misuse, 49
models, 5, 10, 17, 30, 32, 35, 37, 47, 51, 53, 60, 64, 88, 90, 94, 98, 99, 136
moisture, 34, 42, 43, 79, 82, 86, 96, 109, 138
museums, 138

N

NAS, 54
National Academy of Sciences, 54
National Aeronautics and Space Administration, 24, 25, 57, 63, 65, 83, 134, 136
National Park Service, 134
national policy, 61
National Research Council, 14, 65
national security, vii, 6, 13, 17, 46, 58, 64, 67, 70
National Strategy, i, iii, v, 1, 2, 4, 5, 9, 12, 17, 25, 26, 45, 55, 57, 58, 59, 63, 68, 70, 71, 72, 73, 75, 87, 136, 138
natural disaster(s), 27, 31, 81
natural hazards, 71, 93
natural resource management, 78
natural resources, 1, 27, 71, 86
Natural Resources Conservation Service, 134
natural science(s), 4
navigation system, 39
NCA, 15, 57
NOAA, 47, 57, 60, 62, 63, 65, 66, 83, 86, 89, 103, 104, 105, 106, 107, 108, 134, 138
nodes, 28
NPS, 134
NRCS, 134
nutrient(s), 27, 34, 41, 100

O

Obama, 55, 97
Obama Administration, 55, 97
objectivity, 49, 50
ocean buoys, vii, 46
oceans, vii, 4, 33, 36, 41, 43, 73, 135
Office of Management and Budget, 13, 25, 56, 64, 65, 66, 134, 137
officials, 57, 62
oil, 31, 32, 33, 47, 96, 105
oil spill, 31, 33, 47, 105
OMB, 13, 20, 25, 46, 50, 56, 64, 134, 137
openness, 2, 12, 16, 17, 23, 45
operations, 13, 24, 40, 45, 46, 48, 56, 57, 65, 98, 105, 106, 136, 137
opportunities, 10, 13, 16, 28, 92
orbit, 59, 104
overlap, 43
oversight, 2, 9, 17, 136
ownership, 13, 95
ozone, 30, 79, 81, 96, 104, 107

P

pathogens, 27, 36, 37, 41, 101
pathways, 100
periodicity, 100
permafrost, 31, 81
permit, 46, 60
pests, 27
pH, 81
photographs, 101
planning decisions, 35
plants, 27
platform, 6, 17, 20, 46, 86, 89, 98, 101
polar, 95
policy, 2, 4, 6, 8, 20, 34, 46, 50, 51, 59, 60, 61, 65, 66, 69, 72, 77, 90, 132, 137, 138
pollutants, 36, 80
pollution, 30, 31, 36, 37, 38, 99, 105, 109
population, 26, 29, 30, 34, 37, 38
population growth, 37, 38
portfolio, 5, 8, 9, 10, 68, 69, 72, 77, 82, 83, 87, 96, 106, 131
portfolio assessment, 8, 9
potential benefits, 13
power generation, 41
power lines, 32
precipitation, 30, 34, 41, 42, 43, 79, 81, 82, 86, 99, 109
prediction models, 99
preparation, 32, 42, 43, 56
preparedness, 79, 97
preservation, 14, 21, 23, 28, 46, 52, 78, 89
President, 2, 8, 12, 13, 14, 16, 25, 55, 57, 61, 63, 64, 69, 70, 75, 83, 84, 86, 87, 90, 96, 103, 110, 137, 138
primary function, 58
principles, 5, 12, 14, 16, 18, 20, 21, 46, 50, 61, 62, 71, 90, 95, 138
private sector, 5, 13, 88
probability, 30, 49, 101
procurement, 65, 136
producers, 26
programming, 52, 56
project, 22, 54, 101
proliferation, 54
prosperity, 58
protection, 2, 7, 8, 17, 41, 74, 99, 106
public goods, 77
public health, 106
public interest, 68, 76, 78, 80, 137
public investment, 95
public safety, 78, 84
public sector, 77
public service, 67, 68, 69, 76, 77, 78, 79, 80, 81, 82, 84, 97, 98, 102, 103, 137, 138
public-private partnerships, 13
publishing, 20, 22

Q

quality assurance, 89
quality control, 13, 46, 47, 50, 94
quality standards, 49

R

radar, 79, 92, 102, 108, 138
radiation, 81, 86, 108, 135
radio, 38, 106, 107
rainfall, 37, 41
rangeland, 26
rare earth elements, 34
real time, 32
recognition, 62, 83
recommendations, 8, 9, 10, 14, 56, 57, 98, 102
recovery, 4, 28, 33
recreation, 37, 41, 102
recreational, 41
recycling, 43
redundancy, 9, 95
reference system, 48
reform, 51, 65
regulations, 15
reliability, 40
remediation, 27
remote sensing, 9, 28, 35, 60, 62
remotely operated vehicles, 92
renewable energy, 3, 34, 42

renewable energy technologies, 3
repair, 94
replication, 60
reproduction, 60
requirements, 5, 8, 12, 13, 14, 17, 20, 21, 24, 26, 33, 42, 43, 45, 46, 48, 49, 50, 51, 54, 59, 64, 85, 88, 92, 94, 104, 136
researchers, 26, 60, 61
resilience, 38, 75, 97
resolution, 43, 85, 87, 93, 97, 98, 105
resource availability, 5
resource management, 3, 7, 41, 74, 79
resource utilization, 29, 38
resources, 4, 7, 9, 13, 21, 22, 23, 24, 26, 31, 33, 34, 35, 37, 38, 41, 42, 50, 52, 63, 68, 73, 74, 85, 89, 92, 95, 101, 103, 108, 135
response, 1, 3, 4, 24, 32, 42, 43, 68, 71, 79
responsiveness, 6
restoration, 4, 38
restrictions, 93
rights, 42
risk(s), 3, 9, 10, 13, 29, 31, 32, 37, 79, 82, 88, 91, 92, 98, 103, 109
risk assessment, 29
risk profile, 92
runoff, 29, 38, 82

S

safety, 7, 38, 40, 74, 99, 105
salinity, 81
satellites, vii, 7, 33, 35, 39, 46, 64, 74, 79, 82, 84, 89, 91, 94, 100, 103, 104, 105, 106, 107, 108, 109, 113, 130
savings, 87, 91, 92, 93, 96
scattering, 81
schema, 22, 48
science, 3, 4, 9, 16, 29, 32, 51, 59, 60, 66, 82, 83, 85, 86, 88, 89, 96, 97, 103, 131, 138
scientific investigations, 46
scientific understanding, 17, 25, 45
scope, 15, 17, 54, 55, 132
sea level, 30, 81, 96, 105
seafood, 36, 38
sea-level, 37
sea-level rise, 37
seasonality, 30
Secretary of Commerce, 103, 104, 105, 106, 107, 108
security, 2, 6, 13, 39, 49, 64, 78, 79, 84, 95, 132, 136
sediments, 41
semantics, 47, 51
sensing, 4, 64, 65, 73, 84, 101, 135, 138
sensitivity, 81
sensor network, 82, 86
sensors, 4, 6, 45, 46, 48, 51, 73, 79, 82, 89, 94, 102
servers, 52
service provider, 5
services, vii, 10, 12, 15, 20, 21, 22, 23, 34, 38, 43, 45, 47, 50, 51, 56, 58, 67, 68, 70, 75, 76, 81, 82, 84, 86, 87, 88, 90, 91, 93, 94, 95, 97, 99, 102, 131, 132, 137
shape, 43, 59
shelter, 32
shoreline, 98
signals, 39
simulation, 31
social costs, 3
society, vii, 3, 5, 8, 12, 45, 58, 75, 78, 81, 95, 132
software, 21, 22, 24, 54, 93
soil erosion, 27
South Africa, 58, 95
space technology, 39
specialists, 29
specialization, 23
species, 27, 28, 29, 33, 38, 86, 102, 138
specifications, 47
spectrum allocation, 89
SS, 107
stability, 42, 45
stakeholders, 1, 13, 44, 70, 71, 76, 86, 87, 110, 137
standardization, 14, 90
state(s), 10, 27, 40, 64, 79, 81, 97, 103, 137
statistics, 52
stock, 78

storage, 21, 41, 49, 51, 52, 53, 91
storms, 31, 32, 39, 42, 79, 80, 82, 93
stream gauges, vii
stress, 32, 81
stressors, 36, 37, 101
structure, 23, 47, 102, 107
sulfur, 41
sulfur dioxide, 41
Sun, 81
support services, 90
suppression, 78, 79
surveillance, 39, 80
sustainability, 1, 2, 4, 7, 12, 21, 29, 34, 71, 73

T

tanks, 32
target, 73
Task Force, v, 1, 9, 24, 25, 26, 55, 57, 58, 68, 71, 134, 136
taxpayers, 131
teams, 9, 10, 98, 102, 110, 131
technical support, 95
techniques, 9, 24, 48, 51, 82, 83, 86, 90, 93, 102
technological advances, 84
technologies, 6, 10, 11, 12, 25, 35, 39, 40, 52, 82, 94, 102
technology, 4, 22, 23, 35, 39, 82, 92, 93, 104, 106, 138
temperature, 31, 34, 37, 41, 42, 81, 86, 99, 100
temporal variation, 35
terrestrial ecosystems, 33
terrorism, 27
testing, 102, 137
threats, 37, 45
tides, 35
time frame, 35
tornadoes, 28, 31
tourism, 41
trade, 34, 52
trade-off, 53
trajectory, 94
transformation, 15
transmission, 34, 52, 89
transparency, 26, 58
transport, 36, 41, 43, 88
transportation, 7, 34, 35, 40, 41, 74, 78, 79, 94
transportation infrastructure, 40
tropical storms, 42
tundra, 33
turbulence, 43

U

U.S. Department of Commerce, 49
U.S. economy, 70
understanding source, 81
uniform, 15, 90
United, v, 14, 16, 20, 28, 35, 38, 41, 56, 57, 58, 59, 61, 64, 65, 67, 68, 70, 83, 87, 95, 99, 100, 105, 134, 135, 138
United Nations, 138
United Nations Framework Convention on Climate Change, 138
United States, v, 14, 16, 20, 28, 35, 38, 41, 56, 57, 58, 59, 61, 64, 65, 67, 68, 70, 83, 87, 95, 99, 100, 105, 134, 135, 138
urban, 27, 31, 34, 35, 36, 43
urbanization, 28, 37
USDA, 83, 135
USGS, 46, 57, 60, 62, 64, 83, 89, 103, 104, 135, 138

V

validation, 35, 49, 60, 65, 86, 93
variables, 86, 100
variations, 30
vector, 80
vegetation, 28, 34, 109
vehicles, 82, 102
vessels, 92, 101
vision, 58
vocabulary, 22

W

warning systems, 36
Washington, 62, 136
waste, 23
water, vii, 3, 7, 16, 27, 29, 30, 33, 34, 35, 36, 41, 42, 71, 74, 78, 79, 80, 82, 85, 96, 99, 100, 102, 103, 104, 105, 107, 108, 109, 138
water quality, 29, 41, 78, 80
water resources, vii, 104
water vapor, 107
watershed, 4, 30, 41, 100
waterways, 41
weather patterns, 55
web, 21, 46, 47, 48, 50, 52, 66, 91, 137
web service, 21, 46, 47, 48, 50, 52
well-being, 3, 7, 28, 36, 37, 55, 74
wetlands, 101
wildlife, 34, 80
wind power, 34
wind speed, 34, 100, 105
wind turbines, 34
woodland, 101
workers, 55
working groups, 110

X

XML, 48

Y

yield, 18, 26, 28, 41, 69, 77, 83